Changing Attitudes Toward American Technology

THOMAS PARKE HUGHES, Editor
University of Pennsylvania

Changing Attitudes Toward American Technology

Harper & Row, Publishers
New York, Evanston, San Francisco, London

Sponsoring Editor: John Greenman
Project Editor: Karla B. Philip
Designer: Jared Pratt
Production Supervisor: Will C. Jomarrón

CHANGING ATTITUDES
TOWARD AMERICAN TECHNOLOGY

Library of Congress Cataloging in Publication Data

Hughes, Thomas Parke, comp.
Changing attitudes toward American technology.

CONTENTS: Temko, A. Which guide to the Promised Land: Fuller or Mumford?—Roszak, T. Technocracy.—Commoner, B. Technology and the natural environment. [etc.]
1. Technology—United States. I. Title.
T21.H8 301.24'3'0973 74-28920
ISBN 0-06-042983-6

CONTENTS

PREFACE

College students of the late sixties and early seventies know well—even contributed to—the sharp reaction against American technology. This attitude was and is interesting in itself, but the interest is heightened for an older generation that recalls the enthusiastic commitment to technology that was widespread only decades ago. Those familiar with American history find an even more pronounced contrast if they compare the popular sentiments of the present decade with those of a broad section of Americans from decades past. The reader will find, for instance, that World War I and the Great Depression brought doubts about technology to many men and women long before the present generation raised its voice in protest. America's critics, literary and philosophical, have long protested technological advancement, but the very fervor of their protest arose in part from the massive commitment of run-of-the-mill Americans to the machine, the tool, and the technological process.

These readings and essays are intended to document and to interpret changing attitudes toward technology over the past two centuries. The picture that emerges is not a simple contrast between present day rejection and the acceptance of the past, but one depicting the shadings, complexities, and to-and-fro movements, not Whiggish trends unfolding predictably in some progressive—or even retrogressive—fashion.

Attitudes toward technology over a broad sweep of time is a virtually unexplored theme. The specialized articles and books needed to provide the firm base for an authoritative and sweeping interpretation do not exist. Presented here are excellent essays describing and analyzing a class response in one technological period and perhaps a literary reaction in another—but the gaps are wide. Nevertheless, placing a book of readings suggesting tentative interpretations seems justified by the obvious interest of the college generation in the subject and by the conviction that a healthy skepticism will lead them to test the hypotheses.

Faced by the scarcity of secondary materials, I have read widely in the sources to formulate the main themes expounded by selection and interpretation. Having decided to focus upon "middle-brow" in contrast to "high-brow" literature, my reading was principally in the popular periodicals of taste and style. Although each major theme, as reflected in the part titles, is illustrated by only a few articles, the themes themselves emerged from numerous essays read in order to select those published here. For instance, the ambivalence characterizing the articles in Part IV was apparent in numerous articles published in the popular journals of the time.

Relatively, it would have been much less demanding to develop the major themes or attitudes by more numerous short selections, for many essays encountered combined short statements of attitude toward technology with much longer discussions of the technology itself. However, this book is not a *history* of technology and therefore, if these essays had been used, the discussion would have had to be expunged leaving the short statement of attitude as a fragment. Because of the intellectual maturity of the students for whom these readings are intended, I have chosen the route of fewer and longer selections involving sustained argument and providing fuller exposure to the intellectual style of the time. The counter result is fewer witnesses, less diverse evidence.

So that the reader will be able to judge more satisfactorily the validity of my organizing themes, I have, in the introduction, discussed candidly the method by which I selected the major themes or attitudes. In the introduction I have also ventured my explanation for the rise and formation of attitude clusters that seemed to prevail in various periods—enthusiasm, enterprise and power, ambivalence, and so on. As the reader will discover, the attitudes arose more from the way in which the technology was used than from the technology itself. This is not surprising, but the way in which cognition came through the filters of such cataclysmic events as the taming of a wilderness, war, and depression may be. In the conclusion, I have gone beyond short-range time and immediate circumstance in order to interest the reader in persistent as well as contrasting attitudes endemic in America.

Several possible points of confusion that may arise in the reader's mind can be avoided. This book is not about the reaction to science. If it were there would be, for example, selections about the conflict between science and religion. Also, as stated previously, the selections do not sample the opinions of those writing for a scholarly, intellectually esoteric, or professional audience.

The effort has been to capture the sentiments of—as might be anticipated—the intelligent and literate layman.

Having explained that the book is a conscientious search for substance and pattern and in no sense scientific in its samplings, I do not want those who stimulated and helped me in the text's compilation to bear responsibility for its inadequacies. However, I do want to thank them for their contributions. Professor Hugh Davis Graham suggested to me that students should be made aware of contrasts and change in order to help them avoid falling into the error of projecting their mind-set into the past—and future. My wife, Agatha, carefully and critically read the manuscript and improved it obviously; my son, Lucian, used the technology of the copying machine to provide the hundreds of articles that I sampled; and, Mrs. Marthenia Perrin and Mrs. Sylvia Dreyfuss cheerfully and thoughtfully typed the manuscript.

T.P.H.

Changing Attitudes Toward American Technology

INTRODUCTION

There has never been a single "American attitude" toward technology. As the articles in the contemporary section of this book clearly indicate, American attitudes toward technology in the 1960s—especially in the late 1960s—were, at the least, skeptical; at an extreme, hostile and fearful. Other selections, however, show that attitudes have changed and thereby suggest that they will continue to do so. Simply being aware of these shifts should lead the reader to consider that although current attitudes may possibly be ephemeral and certainly may be related to time and circumstance, this does not demonstrate them to be untrue or insubstantial. The final section of this book, a concluding analysis, will raise questions about the validity and substance of many attitudes that have previously been held. If the doubts about past attitudes are convincingly argued, then a constructive skepticism about current ones should follow.

At this point, however, our objective is to delineate the spectrum of attitudes in the past by taking a sample of writings on the general level. The selections, with a few exceptions, were written for nonspecialist readers, those interested in the subject because of their desire to understand rather than to learn. Most of the articles, therefore, are from periodicals like *The Nation, The American Scholar, North American Review, Atlantic Monthly, The New York Times Magazine, The Survey, The Literary Digest, The New Republic,* and *The Living Age.* If an essay is not from such a periodical, it is by an author writing for a similar audience. Even the essay from *Scientific American* is nontechnical and nonprofessional.[1]

The selections are not intended to cover—even on a single level—the broad span of American history and to establish an unbroken chain of fluctuating attitudes toward technology. Rather, they are a sampling of opinion in four clusters: the first section is from the recent past (the 1960s); the second from the decades 1830–1860; the third is taken from the period of about two decades culminating in World War I; the fourth, from the postwar decade and the Depression era.

Each section of readings includes articles that express related attitudes. The common characteristics of the articles within a section result from their being reactions to a major episode in the history of the evolving relationship between American society and technology. For example, in the 1960s—the time frame of Part I—a convergence of trends and events revealed sharply and clearly the "technological forces of the man-made world" and their great impact upon the course of history. In Part II the articles express an affinity arising from all the authors having witnessed technology taming a wilderness, binding a nation, and making possible a life free from back-breaking toil and privation. The general attitude was "the thrill of the technological transformation." In the section covering the period from 1890 up to the entry of the United States into World War I, Part III, the authors share a "realization of technological power"; the power demonstrated by America's rise to the status of the world's leading industrial nation and the power resting to a large extent upon American inventiveness. Part IV is given coherence by the authors having witnessed the terrible destructiveness of technology in World War I and having seen immensely productive technology paradoxically idle in time of want and unemployment during the Great Depression. Some of the selections in Part IV also share the attitude that the response to the doubt and disillusionment should be the exercising of increased control over technology by the professionals who seem to understand it best—scientists and engineers.

Even though the selections that follow are only a sampling, there is evidence that the attitudes they express were representative of opinion about technology during the period in which they were written. That they appeared in general-readership journals indicates broad interest in—or acceptance of—the ideas expressed. Furthermore, the dominant attitudes have been found to be pervasive by historians deeply informed about the various periods. And there were many other articles written about the same time expressing similar attitudes or themes.

The attitudes revealed in Part I should be familiar to any reader of the newspapers and general circulation periodicals of the 1960s. During this decade, concern for the environment increased and reached a peak with the publication of several best sellers about technology and the environment.[2] The concern for the environment focused on technological forces because Americans realized—many for the first time—that man was befouling his nest (to use a catch phrase of the time) and believed that technology was the cause.[3] It dawned upon many Americans that man was literally making a new material world, an artificial one, to replace the natural one being bulldozed away.[4] Furthermore, the new world was judged less habitable than the old, an opinion reinforced by a tendency to romanticize the natural landscape.

It would be difficult to encompass the complex factors and events that would persuasively explain the attention directed to the environment and the concern about—even the hostility toward—technology manifested in the decade of the sixties, but the great smog alerts of the polluted cities and the oil spills of the coastal areas in America, all of which were well publicized by the media, are a partial explanation. The large-scale exploitation of military technology in Vietnam may also have brought some Americans to attach their negative reactions to the war to the highly destructive technology with which the nation waged it. Another stimulus for the formulation of attitudes toward the technological forces of the man-made world was a growing conviction in the 1960s that the techniques, objectives, and values of military and civilian technology and science were threatening to subvert individual will and freedoms.[5] Whatever the causes of the heightened concern about technology, well-articulated attitudes about it dominated the essays that frequently appeared in periodicals and newspapers. A persuasive indication is the number of articles published in leading general circulation journals about environmental problems. The *Readers' Guide* indexes more than a hundred periodlcials under numerous and changing subject headings.[6] Those who prepare the index introduce new subject headings and drop old ones as the interests of the reading public shift. There was a steady rise from 1961 to 1968 in the number of articles on air pollution, a subject category in the *Readers' Guide* (the annual number more than doubled during the period). The number of articles under the heading Oil Pollution of Rivers, Harbors, etc. increased even more dramatically from three to four annually in 1961–1963 to twenty-six in 1967–1968. Also of significance was the introduction of a new subcategory in 1965–1966: Man—Influence on Nature. In 1967–1968, fourteen articles were published in this relatively new category.

In the case of Part II the themes elucidated are in accord with the findings of historians who have probed the ideas—the mind set—of Americans during the first half of the nineteenth century. Outstanding among these historians was the late Perry Miller, whose death cut short his last book, *The Life of the Mind in America.*[7] It has only an unfinished section on, and an outline of, the attitudes he found Americans taking toward technology in the first half of the nineteenth century, but it permits one to see the shape his interpretation was taking as he read the sources expressive of the middle ground of American thought. He mined a vein that was neither high-brow nor popular culture, but the cognition, reflection, and speculation that might be labeled conventional wisdom. In the platitudes and banalities of the inaugural speeches, the Fourth of July orations, and the commencement addresses, he found commonly held attitudes among the literate and articulate. These attitudes he aptly characterized as expressing the "thrill of the technological transformation."[8]

Americans, Miller realized, became increasingly aware of their genius for technology as they fashioned it and used it to subdue a continent. Inventors, artisans, and engineers showed that knowledge was power and that mind prevailed over matter as they seemed to Americans to be fulfilling God's plan by completing his creation.

Another historian, Hugo Meier, working with an intent and style not unlike Miller's, found similar attitudes. In an article entitled, "Technology and Democracy, 1800–1860,"[9] Meier stressed the American belief that technology would provide "the physical means of achieving democratic objectives of political, social and economic equality."[10] The fortunes of technology were linked with those of the new American political and social system. Meier found a faith in progress and an assurance that the evils of British industrialism would not plague the new society.[11]

Leo Marx and Marvin Fisher should also be mentioned as historians whose findings are in accord with the themes in the readings of Part II. Marx, in *The Machine in the Garden,*[12] observed the recurrent image of the machine—especially the locomotive against the landscape, as depicted in early nineteenth-century American literature—and interpreted this image as indicative of the excitement of the technological transformation of a wilderness and of the anticipation that the end would be, not an industrial wasteland, but the "middle landscape" of a garden. Marvin Fisher in *Workshops in the Wilderness*[13] also juxtaposed the man-made and the natural, for he read in the reports of foreign visitors the vivid and lasting impressions of the greying of a greenland.

The themes developed in Part III are those found often in the contemporary sources I have used in writing of inventors and engineers who flourished during the period. For instance, the effort to place the creative work of Thomas A. Edison (1847–1931) and Elmer A. Sperry (1860–1930) within the economic, social, and political context of their times required broad reading in the technology-related literature of this era in America.

Interest in invention and inventors was a manifestation of the realization of the power of technology.[14] This interest ran especially high immediately before World War I. Several decades later, the word and the concept "inventor" were almost submerged by the rise of the industrial research scientist. An indication of the high level of interest before the war was the number of articles about inventions that appeared in general readership journals. During five years from 1910 to 1914, an average of twenty-five articles a year on inventions were listed in the *Readers' Guide to Periodical Literature;* during the quarter-century from 1900 to 1925 the annual average was only about ten. Similarly, there were forty entries for articles on Thomas Edison in the same five-year period (1910–1914) in the *Readers' Guide.* In the prior decade (1900–1910) there were only twenty-five, and in the subsequent decade (1915–1924), thirty-seven. His renown and the general interest in invention were mutually reinforcing.

The selections in Part IV were chosen in a somewhat different way from that used for the other parts of the book. Reading in the periodicals on the level described earlier (*The Nation, New Republic, The Literary Digest,* and *Survey,* for example) and expecting to find an attitude toward technology reflective of the prosperity, the superficial optimism, and the freneticism of the post-World War I decade, I found instead a striking ambiguity toward technology and applied science. The origins of this disenchantment may well have come from England, for influential social critics like Bertrand Russell and J. B. S. Haldane were spearheading a debate about the social effects of applied science and technology that seems to have been deeply permeated by the shock of seeing close at hand the terrible destructivenss of modern military technology.[15]

American commentators, influenced by British attitudes, expressed an ambiguity stemming in part from the violent contrast between the technological euphoria of the nineteenth century ("the thrill of the technological transformation" and the "realization of technological power") and the distant but comprehensible horrors of World War I with its exploitation of science and technology. Americans picked up issues formulated in England, such as the question of whether further scientific and technological advance would impede human welfare and whether military technology would end war or end the race, and explored these from an American perspective. More subtle issues like the influence of the mindless work routine of American production technology upon the workers' sense of meaning and achievement also found a place in American commentary. The ambivalence manifested itself in the raising of questions and the exploration of issues that were ignored in the Pollyanna-ish essays of the nineteenth century.

Some of the doubts raised during the period from 1919 to 1935 are suggested by the titles of articles published in the general readership journals. For instance, in 1919 the readers of *The Literary Digest* were asked to ponder the "Dubious Benefits of Science" and in 1921 to ask, "Is the Public Hostile to Science?" A year earlier, the same journal published an article titled "Science as a Curse to Mankind." *Current Opinion* reflected this growing attitude in the title of an article carried in 1924; "Science Viewed as Man's Destroyer." *Harper's Magazine* wondered in 1928 "Is Science a Blind Alley?" and the *Dial,* two years earlier, "Is Science Superstitious?" In 1926, *Survey* was less tentative in publishing an article titled "Some Sins of Science." *Harper's* posed a question again in 1926—"Will Science Destroy Religion?" By 1934, the same magazine was asking, "Is Science a Fashion of the Times?" *The Literary Digest* wanted to know, "More Science or Less?" No wonder that the *Christian Century* in the same year published an article entitled "Revolt Against Science." During the 1920s and 1930s science was often—if not usually—associated with applied science, and the questions and doubts about science were in fact often directed toward what would commonly be called technology a half-century later.

As would be expected, the disillusionment was heightened after 1929 by the coming of the Great Depression. Not only was technology terribly destructive; it was at the root of a calamitous social maladjustment—unemployment. Even before then, however, some saw a way out through the imposition of social controls on applied science and technology, especially by those who should have understood them best, the scientists and engineers.

This hope for scientism and technocracy is found in the selections in Part IV.[16]

A convincing indicator of the widespread interest in technocracy during the depths of the Depression is the large number of magazine articles published about it between 1932 and 1935. During those four years, the *Readers' Guide* listed sixty-three articles under the rubric "Technocracy." An indexing of more scholarly periodicals, scientific and technical journals, and newspaper articles would have added to the total. The attention directed to technocracy and the attitudes it encompassed was of intense but brief duration: The category Technocracy was not used in the *Readers' Guide* before 1932 and after 1935.[17]

Scientism can be defined as the conviction that the assumptions, methods, and discoveries of science are applicable to social problems. *Technocracy* can be defined generally as the belief that efficiency and production should be society's primary goals and that engineers and technically trained managers using a rational, quantitative, and systematic methodology are best equipped to make the decisions that would realize these objectives. During the Great Depression of the 1930s, however, essential principles of technocracy were incorporated along with more specific postulates in a movement that called itself "Technocracy."

The movement attracted considerable interest and influenced American attitudes toward technology because Technocracy responded to some troublesome paradoxes. The Technocrats insisted that unemployment and want existed alongside idle factories because technology had been mismanaged, not because of an inherent and irremediable characteristic of modern technology. The movement promised a blueprint for social reorganization that would realize the full and beneficent potential of modern technology. The selections in Part IV focusing on Technocracy—defining, supporting, and attacking it—represent postwar American ambivalence heightened by the shock of the Depression. Earlier American optimism was spurred once again by the promise of a technological world in which labor would be light and the man-made environment would be esthetically, spiritually, and materially satisfying.

Depression, wars, and vistas of unexploited land and resources have stimulated reactions or attitudes toward the uses or potential of technology. But what influences these attitudes at any one time is a complex combination of events, trends, and ideas; a war or a depression may be only the most obvious immediate cause of the crystallization of attitudes. Since the purpose of this book is not an exhaustive explanation of attitudes but a sampling and description, this is not the place to explore in depth the question of attitude formulation. Before considering the attitudes highlighted in this book, however, it seems desirable to catalog briefly one class of events that, along with wars, depressions, and much else, certainly helped shape the changing attitudes toward technology in America: the inventions and engineering feats of the developing technology.[18]

Attitudes of the period from 1830 to 1860 were certainly influenced by the appearance of the steamboat, steam locomotive, and telegraph in a vast land needing transportation and communication. The locomotive and the railroad were British inventions but, as early as 1830, Peter Cooper (1791–1883) designed and constructed the Tom Thumb, the first steam locomotive built in America. Construction on the Baltimore and Ohio Railroad began in 1828. By 1853, this line and the rails of the Pennsylvania, the New York Central, and the Erie had linked the Eastern seaboard with the Mississippi Valley through that river or one of its tributaries. Altogether, more than 30,000 miles of railroad had been laid in the United States by 1860.[19]

Steamboats also greatly impressed Americans and convinced them that technology would overcome the vast distances of the new world. Unlike the locomotive, the steamboat could be called an American invention because of the pioneer inventions and developments of John Fitch (1743–1789), James Rumsey (1743–1792), and Robert Fulton (1765–1815). By 1855 there were more than 700 American steamboats, many of a unique western design, plying the rivers of the Mississippi Valley.[20] After Samuel F. B. Morse (1791–1872) laid his telegraph line from Baltimore, Maryland, to Washington, D.C., in 1844, the expanding telegraph network provided additional evidence that the loneliness and silence of the countryside and the isolation of towns would pass.[21] As one writer observed in 1855, "in no important piece of intelligence is the country—east, west, north, or south—more than a few hours, seldom more than a few minutes, behind the metropolis."

As the selections of Part II also show, labor-saving machinery stimulated reflection and speculation. Although stationary steam engines were slowly spreading throughout the United States during the first half of the nineteenth century, water-powered mills, fulfilling a wide variety of functions through the machinery they drove, were more abundant in a country where water power sites had not been virtually exhausted, as in England.[22] As an American wrote in 1832, "we arrest the water as it flows onward to the ocean. It must do so much spinning, so much weaving, or so much grinding, before it can be allowed to pass on." By the mid-nineteenth century, labor-saving and resource-intensive machines were representative of one characteristic of American technology. In half a century, the United States had become a leading developer of machine tools and production machines, and a nation known for its system of quantity production.[23]

By the end of the century, inventions and the act of invention were central concerns of those interested in the development of technology, as the selections in Part III suggest. In preceding selections, technology was seen as man compelling the forces of nature to do his work and to extend his facility for communication and transportation; but by the turn of the century, the resourceful inventor and engineer had convinced many Americans that technology was a broader, generalized, man-made force that could be applied at will to a wide variety of problems as they arose. Technology could bring order out of chaos, provide boundless energy, support business enterprise, and win wars.

Thomas A. Edison was as responsible for this attitude as one man and his work could be. He turned his inventive genius to a number of problems—the multiple telegraph, the incandescent lamp, the electric generator, the telephone, the phonograph, the electric battery, and the magnetic ore separator, for instance—as need and personal interest moved him. If Edison's inexhaustible creativity was not sufficient evidence of the potential and multifarious power of technology, then there were the Wright brothers, George Westinghouse, Alexander Graham Bell, Lee De Forest, Elihu Thomson, Charles Brush, Charles Hall, Edward Dean Acheson, Elmer A. Sperry, and many other successful and well-known American inventors who flourished between 1880 and 1920.[24] Not only were Americans impressed by the quality of their inventors, they were impressed by the quantity of American inventions as well. American invention—as patent statistics seemed to show—was by the end of the century "a gigantic tidal wave of human ingenuity and resource, so stupendous in its magnitude, so complex in its diversity, so profound in its thought, so fruitful in its wealth, so beneficent in its results, that the mind is strained and embarrassed in its effort to expand to a full appreciation of it."

The use of the fertile mind of the inventor for the solution of military problems during the armaments race after 1890 and during World War I was another manifestation of the multifaceted power of technology.[25] Edison did not doubt that technology, specifically the resources of the inventive mind, could be directed to the problems of war—if the need existed and the will were so inclined. The war would be won, Edison was convinced, by inventors, engineers, and technicians, not by soldiers. The history of technological development during the war does not contradict him. The machine gun, poison gas, the airplane, the submarine, and the tank resulted in the "technological surprise" that at least one historian believes followed upon the failure of statesmen and soldiers to foresee the role technology would have in 1914 and after; it also brought the horrendous casualties of the war of attrition.[26]

The victory was in part a triumph of technology; to some social critics, it was also a foreboding of disaster. The selections in Part IV are, in part, a reaction to witnessing the metamorphosis of the machines, processes, and utensils of peace into instruments of destruction. In the 1830s Americans had visions of a world man could make possible with locomotives, power looms, steamboats, precision machine tools, textile mills, and the like. The war brought the realization that technology, like Janus, was two-faced—or even, like Hydra, multiheaded. So, in a sense, the attitudes in Part IV are reactions to essentially the same power and many of the same machines and processes about which Americans had been enthusiastic earlier; the reactions now, however, were those of man seeing technology through a transforming filter of historical experience—total technological war.

Several of the selections on technocracy and scientism are also a negative reaction to machines and processes seen through the filter of historical experience—in this instance, the Great Depression. The automobile assembly line, the convenient, available, and plentiful power of large electrical power systems, the rationally organized and highly efficient petroleum refineries and heavy chemical plants, the apparatus and skills that built the subways and skyscrapers of early twentieth-century New York were, after 1930, shut down or only partially utilized.[27] Idle means of production in the midst of human need forced many Americans to see technology (the means of production) as dependent upon the relations of production (business management, financial control, and general social organization).

The themes of Part I stem partly from the consciousness of the hostile environment of the man-made world—the urban blight, noise, pollution, and banal and oppressive architecture. But the alienation and anxiety expressed in these essays is also a reaction to the scale and related momentum of twentieth-century technology. The atom bomb, the monstrous traffic of the automobile, the fantastic acceleration and velocity of jet airplanes and rocket-propelled missiles, the massive and dense communications networks—phenomena familiar to the reader—were only in part under human control, to the extent that contingencies and derivative effects could be foreseen.[28] As the literature on modern technology reveals, however, there was a deep fear that in aggregate and in unforeseen ways technology was "an automobile without driver, steering wheel, or brakes, but crammed with demoralized passengers hurtling full-speed toward doom."

Notes

[1]The *Scientific American,* judging by the character of its articles in the late nineteenth and early twentieth centuries, was directed at an audience made up of engineers and inventors interested in learning of the latest developments in all fields of technology and applied science; it was also read avidly by young Americans who aspired to be engineers or inventors and older Americans who regretted that they were not. *Scientific American*'s pages convey the essence of the excitement of America's realization of the power of technology.

[2]Notable and widely read among the books on contemporary technological society, its problems, and counteraction and counterculture were Theodore Roszak, *The Making of a Counter Culture: Reflections of the Technocratic Society and Its Youthful Opposition* (Garden City, N.Y., 1969) and Charles Reich, *The Greening of America: How the Youth Revolution Is Trying to Make America Liveable* (New York, 1970).

[3]Among those books arousing popular concern about the environment in the 1960s were Rachel Carson, *Silent Spring* (Cambridge, Mass., 1962); Barry Commoner, *Science and Survival* (New York, 1966); Paul R. Ehrlich, *The Population Bomb* (New York, 1968); and Loren Eiseley, *The Firmament of Time* (New York, 1960). For readings, see John Opie, ed., *Americans and Environment* (Lexington, Mass., 1971), and Roderick Nash, ed., *The American Environment* (Reading, Mass., 1968).

[4]In the late sixties, historian Lynn White, Jr., wrote, "concern for the problem of ecological backlash is mounting feverishly." He asked his readers to consider the proposition that the roots of the problem originated in the medieval West, when Christians intensified the exploitation of nature believing

that God willed man to have dominion over the earth. White, "The Historical Roots of Our Ecologic Crisis," *Science,* 155 (March 10, 1967), 1203; or in *Machina Ex Deo: Essays in the Dynamism of Western Culture* (Cambridge, Mass., 1968), pp. 75–94. For penetrating and succinct abstracts of White and other recent articles and books, see *Technology and Social History,* prepared by John Weiss, Research Review no. 8 of the Harvard University Program on Technology and Society (Cambridge, Mass., 1971). Leo Marx has provided a historian's background for the contemporary ecological crisis in "American Institutions and Ecological Ideals," *Science,* 170 (November 27, 1970), 945–952.

[5]Selections from the sixties emphasizing the overwhelming influence of technology on the individual and society will be found in the first section of readings that follow. Others developing the same or similar themes were Herbert Marcuse, *One-Dimensional Man* (Boston, Mass., 1964); Marshall McLuhan, *Understanding Media: The Extensions of Man* (New York, 1964); Victor Ferkiss, *Technological Man: The Myth and the Reality* (New York, 1969); and Jacques Ellul, *The Technological Society,* trans. John Wilkinson (New York, 1964). Among the numerous articles of note are John McDermott, "Technology: The Opiate of the Intellectuals," *The New York Review of Books,* July 31, 1969, and Paul Goodman, "The Human Uses of Science," *Commentary,* 30 (December 1960), 461–472. Philosophical reactions to the momentum of technology vary across a spectrum from fatalistic resignation to faith in the possibility of control by men of good will, intelligence, experience, and action. For selections from "Philosophers of the Technological Age," see Albert H. Teich, ed., *Technology and Man's Future* (New York, 1972). See also an extensive annotated bibliography, "Bibliography of the Philosophy of Technology," Carl Mitcham and Robert Mackey, *Technology and Culture,* 14; 2, part II (April 1973), i–xv, sl–s205.

[6]*Readers' Guide to Periodical Literature: An Author and Subject Index* (New York, 1901 to present). It indexes general and popular periodicals.

[7]*The Life of the Mind in America from the Revolution to the Civil War* (New York, 1965). For a critical analysis of Miller with references to his works, see David A. Hollinger, "Perry Miller and Philosophical History," *History and Theory,* 7 (1968), 189–202.

[8]For more on the technological transformation as it pertained to the wilderness, see Roderick Nash, *Wilderness and the American Mind* (New Haven, 1967), a study focusing upon conservation, and Hans Huth, *Nature and the American: Three Centuries of Changing Attitudes* (Berkeley, 1957).

[9]*Mississippi Valley Historical Review, 43* (March 1957), 618–640.

[10]A major democratic objective was lifting the burden of onerous labor from everyman's back. For an extended essay on American technology's labor-saving characteristic, see Hrothgar John Habakuk, *American and British Technology in the Nineteenth Century: The Search for Labour-saving Inventions* (Cambridge, England, 1967).

[11]In the first half of the nineteenth century, American factory pioneers professed a determination to make America's factories a morally elevating influence, and many factory owners acted in accord. Charles L. Sanford, "The Intellectual Origins and New-Worldliness of American Industry," *Journal of Economic History,* 18 (1958), 1–16. This is one of the essays summarized in John Weiss, *Technology and Social History.*

[12]*The Machine in the Garden: Technology and the Pastoral Ideal in America* (London, 1964).

[13]*Workshop in the Wilderness: The European Response to American Industrialization, 1830–1860* (London, 1967). Marvin Fisher has also written, in a similar vein, "The Iconography of Industrialism," *American Quarterly*, 13 (1961), 347–364.

[14]Thomas Parke Hughes, *Elmer Sperry: Inventor and Engineer* (Baltimore, 1971), and forthcoming (1975) study of Edison to be published by the Science Museum (London). Also see essays on inventors and engineers of this period by Thomas P. Hughes in *Dictionary of Scientific Biography* (C. M. Hall, T. Midgley) and *Dictionary of American Biography* (V. Bendix, C. A. Stone and E. S. Webster, F. Cottrell, and F. B. Jewett). An indication of how large invention and inventors loomed on the American horizon is that in the two decades following the St. Louis Exposition of 1904, a year that marked the twenty-fifth anniversary of Edison's invention of the incandescent lamp, Edison was often chosen as America's "greatest" or "most useful" citizen in popular newspaper or magazine polls. Matthew Josephson, *Edison: A Biography* (New York, 1959), p. 434. Also on Edison, see F. L. Dyer, T. C. Martin, and W. H. Meadowcroft, *Edison, His Life and Inventions*, 2 vols. (New York, 1929).

[15]See pp. 224–238 of this text for the Russell position on technology, science, and society; the more sanguine J. B. S. Haldane attitude will be found in "Science and Ethics," *Harper's Magazine*, 157 (June 1928), 1–10.

[16]Ronald C. Tobey, *The American Ideology of National Science, 1919–1930* (Pittsburgh, 1971), considers the efforts of scientists, like Robert A. Millikan, to formulate and proselytize an ideology that would win broad national support for science and its institutions.

[17]Secondary works on technocracy include Henry Elsner, Jr., *The Technocrats: Prophets of Automation* (Syracuse, N.Y., 1967). To better understand the post-World War I Technocracy movement, the reader should consider its roots, which extended into scientific management or Taylorism. Some of the followers of Frederick W. Taylor, the leading exponent of scientific management about the turn of the century, insisted that its principles and methodology could be applied to general social reform, an attitude similar to that of the Technocrats. On Taylorism, see Hugh G. J. Aitken, *Taylorism at the Watertown Arsenal* (Cambridge, Mass., 1960); on scientific management broadly conceived, see Samuel Haber, *Efficiency and Uplift: Scientific Management in the Progressive Era, 1890–1920* (Chicago, 1964). On engineers' conviction—or lack of conviction—that their world view and methodology merited them more status and influence, see Edwin T. Layton, Jr., *The Revolt of the Engineers: Social Responsibility and the American Engineering Profession* (Cleveland, 1971).

[18]There is no standard history of American technology from its beginnings to the present. A history placing evolving American technology in a context including economic, scientific, and political factors as well as sociological and psychological influences is obviously needed. In the meantime, the following pioneer works are extremely helpful: *Technology in Western Civilization*, Melvin Kranzberg and Carroll W. Pursell, Jr., eds., 2 vols. (New York 1967) (volume two pertains especially to the United States; both volumes have substantial bibliographies); *Readings in Technology and American Life*, Carroll W. Pursell, Jr., ed. (New York, 1969), also with a bibliography; *Technology and Social Change in America*, Edwin Layton, Jr., ed. (New York, 1973) has recent articles; Brooke Hindle, *Technology in Early America: Needs and Opportunities for Study* (Chapel Hill, N.C., 1966) has an introductory essay followed by a lengthy bibliographical essay on nineteenth-century American technology; Roger Burlin-

game, *March of the Iron Men: A Social History of Union Through Invention* (New York, 1938) and his *Backgrounds of Power* (New York, 1949); John W. Oliver, *History of American Technology* (New York, 1956); Dirk J. Struik, *Yankee Science in the Making* (New York, 1962); *Science and Society in the United States,* David D. Van Tassel and Michael G. Hall, eds. (Homewood, Ill., 1966); John A. Kouwenhoven, *Made in America: The Arts in Modern Civilization* (New York, 1948). Portions of the following histories of engineering deal with the United States: Richard S. Kirby, S. Withington, A. B. Darling, and F. G. Kilgour, *Engineering in History* (New York, 1956), and James Kip Finch, *The Story of Engineering* (Garden City, N.Y., 1960).

Eugene S. Ferguson's *Bibliography of the History of Technology* (Cambridge, Mass., 1968) is the standard reference, but it is not limited to American history. For bibliography, see also "Current Bibliography in the History of Technology," published annually in *Technology and Culture* and compiled by Jack Goodwin. *Technology and Culture* is the quarterly of the Society for the History of Technology. Articles on American technology also appear in *ISIS, Business History Review,* and the *Journal of Economic History.*

[19]In addition to informative histories stressing the economic history of American transportation, such as George R. Taylor, *The Transportation Revolution, 1815–1860* (New York, 1951), there is a technological monograph, John H. White, Jr., *American Locomotives: An Engineering History, 1830–1880* (Baltimore, 1968).

[20]Like White's *American Locomotives,* Louis C. Hunter's, *Steamboats on the Western Rivers: An Economic and Technological History* (Cambridge, Mass., 1949) explores the technological core. He moves on to consider related non-technological factors.

[21]There are accounts of the telegraph focusing upon its economic, business, and biographical aspects, but unfortunately no technological history like the White and Hunter books noted above.

[22]Carroll W. Pursell, Jr., in *Early Stationary Steam Engines in America* (Washington, D.C., 1969) considers the diffusion of the steam engine to—and its application in—the United States. On the persistent use of water power, see Peter Temin, "Steam and Waterpower in the Early Nineteenth Century," *Journal of Economic History,* 26 (June 1966), 187–205.

[23]*The American System of Manufactures; the Report of the Committee on the Machinery of the United States, 1855 . . . ,* ed., with an introduction, by Nathan Rosenberg (Edinburgh, 1969).

[24]On American inventors who flourished between the Civil War and World War I, see Matthew Josephson, *Edison: A Biography* (New York, 1959); F. E. Leupp, *George Westinghouse: His Life and Achievements* (Boston, 1918); Robert V. Bruce, *Bell: Alexander Graham Bell and the Conquest of Solitude* (Boston, 1973); David Woodbury, *Beloved Scientist: Elihu Thomson* (New York, 1944); Junius Edwards, *The Immortal Woodshed: The Story of the Inventor Who Brought Aluminum to America* (New York, 1955); Fred C. Kelly, *The Wright Brothers* (New York, 1943); *The Papers of Wilbur and Orville Wright,* Marvin W. McFarland, ed., 2 vols. (New York, 1953); Lee De Forest, *Father of Radio: Autobiography* (Chicago, 1950); Michael I. Pupin, *From Immigrant to Inventor* (New York, 1923); John W. Hammond, *Charles Proteus Steinmetz: A Biography* (New York, 1924); and Thomas C. Martin, *The Inventions, Researches, and Writings of Nikola Tesla* (New York, 1894).

[25]Elting Morison in his *Men, Machines, and Modern Times* has several case studies of invention and engineering for the military during the late nineteenth century. See especially "Gunfire at Sea: A Case Study of Innovation," pp. 17–44. Morison's *Admiral Sims and the Modern American Navy* (Boston, 1942) also provides information about and insight into the relationship between expanding technology and modern military power. Elmer Sperry was considered by the U.S. Navy as the inventor who had contributed most to its strength before and during World War I (Hughes, *Sperry*). Also on the U.S. Navy, the pre-World War I armaments race, and technology, see a series of five articles by J. Bernard Walker in the *Scientific American,* 105 (1911).

[26]Raymond Aron in *The Century of Total War* (Boston, 1955) presented the thesis of "technological surprise," but historical studies of various American technological developments and their impact upon World War I strategy and tactics, like I. B. Holley, Jr., *Ideas and Weapons: Exploitation of the Aerial Weapon by the United States During World War I* (New Haven, 1953), are needed. Older studies include Frank Parker Stockbridge, *Yankee Ingenuity in the War* (New York, 1920).

[27]On the development of mass production techniques by Henry Ford and his associates, see Allan Nevins, with Frank E. Hill, *Ford, the Times, the Man, the Company . . . 1863–1915* (New York, 1954), and *Ford: Expansion and Challenge, 1915–1933* (New York, 1957). Also see John Rae, *The American Automobile* (Chicago, 1965), and James T. Flink, *America Adopts the Automobile, 1895–1910* (Cambridge, Mass., 1970). Harold C. Passer's *The Electrical Manufacturers, 1875–1900: A Study in Competition, Entrepreneurship, Technical Change, and Economic Growth* (Cambridge, Mass., 1953) pertains to early manufacturing, but a history of electrical utilities has not yet been written. However, Forrest McDonald's *Insull* (Chicago, 1962) not only presents the life of a utility magnate greatly affected by the Great Depression, but also explains the technological changes with which he was involved. On petroleum refining in the age of the automobile, see John Enos, *Petroleum Progress and Profits: A History of Process Innovation* (Cambridge, Mass., 1962), and Harold F. Williamson et al., *The American Petroleum Industry: the Age of Energy, 1899–1959* (Evanston, Ill., 1963).

[28]For a history of the Manhattan Project (development of the atom bomb), see Richard G. Hewlett and Oscar E. Anderson, *The New World, 1939–1946: A History of the United States Atomic Energy Commission,* vol. I (University Park, Pa., 1962). On the development of radio—not its recent social impact—see W. Rupert Maclaurin, *Invention and Innovation in the Radio Industry* (New York, 1949). For the business and technological development of the airplane, see John B. Rae, *Climb to Greatness: The American Aircraft Industry, 1920–1960* (Cambridge, Mass., 1968). On rockets, see *The History of Rocket Technology,* Eugene Emme, ed. (Detroit, 1964). On the increased scale of technology and the growing role of government, see A. Hunter Dupree, *Science in the Federal Government: A History of Policies and Activities to 1940* (Cambridge, Mass., 1957). Government patronage and technological development during World War II is discussed in James Phinney Baxter, III, *Scientists Against Time* (Cambridge, Mass., 1946). On industrial research, see Kendall Birr, *Pioneering in Industrial Research: The Story of the General Electric Research Laboratory* (Washington, D.C., 1957).

PART I

The Shock of Realization: The Man-made World

INTRODUCTORY NOTE

Allan Temko's analysis of the attitudes of Lewis Mumford and Richard Buckminster Fuller toward technology, "Which Guide to the Promised Land: Fuller or Mumford?" summarizes well the contrast between the deep doubt—even pessimism—welling up in the late 1960s (Mumford) and the persistent optimism toward the technological transformation (Fuller). Reduced to simple concepts and terms, the ideas of both men captured the interest of a broad segment of the public. Their books, articles, and television appearances caused considerable discussion among those interested in the long-term problems of contemporary Western society.

As Allen Temko emphasized in the essay that follows, Mumford and Fuller, although sharing a concern about technology, differed dramatically in their recommendations about how Americans should use it. Fuller's attitude, like that of many Americans before him, was that of a visionary longing for a world transformed to an even greater extent by new technology. He was not deeply alarmed at the prospect of a society ordered systematically in accord with models or analogies drawn from the machines and processes that man himself had made. He urged a "total use of total technology for total population."

Fuller was convinced that the organization of technologically structured societal systems would result in the fulfillment of the human potential. Mumford was not. Mumford was alarmed by the trend in modern times toward the imposition of rational systems upon life, for he saw these systems, modeled upon machines, as antithetical to the biological order and harmony of life. One of the systems that Mumford detested because of its mechanization of life was utilitarianism which, in twentieth-century garb, he identified as technocracy and scientism with their "crass mechanistic notions of 'efficiency.' " Within these systems, men became components whose essential character was denied by mechanistic imperatives. Men, Mumford believed, fulfill themselves through awareness of their capacity for "ritual, art, poesy, drama, music, dance, philosophy, science, myth, religion."

Allan Temko, an architectural critic, is the author of *Eero Saarinen* (New York, 1962), a volume in the Makers of Contemporary Architecture Series and of *Notre-Dame of Paris* (New York, 1959).

Which Guide to the Promised Land: Fuller or Mumford?

ALLAN TEMKO

They both understand the crisis; they both know all the facts; they are both brilliant and thoughtful men. It is a measure of our ecological crisis today that they disagree utterly on where to go from here.

No matter how remotely separated Lewis Mumford and Richard Buckminster Fuller may be in temperament and thought—which is as far apart as ultraviolet and infrared in the diffused spectrum of modern environmental theory—they stand very close on several counts. Both believe that man's potentialities are virtually limitless, but that he has failed to make the most of them, or else has grievously misused them, so that technological civilization—which if rationally organized could transform the world into a heaven on earth—is in fact in a parlous way and needs a thorough overhaul on a global basis. Therefore they recommend, each from a radically different point of view but with equal fervor, that the chief components of the world environment, such as large cities, be renewed—not to say totally rebuilt—according to "comprehensive" and "integrated" design. Whatever else they may be, Mumford and Fuller are *integrators* who seek wholeness in a new order of man in an age of widespread cultural and physical disintegration.

Both of these "one-world" men believe, moreover, that before a valid global order can be established, man's spiritual and intellectual relationships with his physical environment, including of course machine-made environment as well as natural surroundings, must be profoundly redefined and revised. To Mumford the humanist, as to Fuller the technologist, man's relation to the machine itself is of supreme importance. Indeed, in an age of intercontinental missiles that has already witnessed genocidal exterminations, that is afflicted by hunger, ill-health, drudgery, ignorance, and squalor for hundreds of millions of people, not only the welfare but the very survival of civilization hinges on the humane use of depersonalized technology. Consequently Mumford and Fuller insist that the machine must be put more fully at the service of man, although their diagnoses of man's technological dilemma, together with their prescriptions for sweeping remedial action, differ so markedly that it would seem they are grappling with utterly separate problems.

Yet both Mumford, the spokesman for man in all his human complexity, and Fuller, who enthrones technics as the single solution to the difficulties of mankind, are nothing if not "generalists." Each, with some pride, describes himself as such and disdains slavish specialization as a bane of the present age. Even when they address themselves to comparatively narrow subjects—Mumford to the neolithic village, perhaps, or to the psychology of Jung, or to gardening at his home in Dutchess County; and Fuller to the etymology of the word "dome," to sailing off the Maine coast, or the design of drainage systems—each of them thinks within the frame of a much larger conceptual scheme. The hyphen is a formidable device in their intellectual arsenals, enabling them to interconnect a bewildering range of disciplines within their broad, unifying philosophies.

Such catholicity scarcely endears them to conventional pedants; and for their part Mumford and Fuller, who are otherwise kindly men, have little patience with conventional wisdom, whether exhibited by professors, politicians, or plumbers. Neither of them, it goes without saying, is an academician, although both have held famous professorships and have spent much time in universities. (Fuller, who never got past a tumultuous freshman year at Harvard, is actually a research professor at Southern Illinois University at Carbondale, but he is seldom there; whereas Mumford, whose studies at City College, Columbia, the New School for Social Research, and the museums and libraries of New York were so free that his only academic degrees are honorary, was installed last year as scholar in residence at Leverett House at Harvard.)

On the campus, understandably enough, their all-encompassing doctrines often evoke the private gibes of pedagogues who resent their brilliant forays into carefully fenced scholarly terrain, but such professorial skepticism does not dampen the unreserved enthusiasm of students who are "turned on" by these vigorous and fearless septuagenarians. For even though their styles could not be more different—Mumford's measured eloquence and smiling dignity contrasting sharply with Fuller's bombastic and garrulous monologues—they are both inspiring and very great teachers who, like Henry Adams and H. G. Wells before them, fear that time is running out and that the fate of civilization may soon be decided, in Wells's phrase, by "a race between education and catastrophe."

Finally, and not least important, Mumford and Fuller are unmistakably Americans. Both were born in 1895, Fuller in Massachusetts, and Mumford in New York. The differences in their backgrounds are significant. Fuller, the descendant of generations of Harvard men, whose forbears came to this country in 1630, attended a Milton Academy that was far removed from the Stuyvessant High where Mumford, whose old New York family had come to know "genteel poverty," learned at firsthand the inhaustible variety and vitality—and the crushing realities—of the melting pot of the modern industrial metropolis. Fuller was exposed to comparable experience later, at the nadir of his career in the 1920's, when he lived for a time in a Chicago slum. Yet he never came to regard the city, as Mumford has, as a "time-structure" that itself is a formative factor in civilization. Fuller, the inventor, has sought to create an utterly new urban environment. Mumford, the historian, has seen the city as an institution that is the expression of social as well as physical forces—"history made visible"—and that transforms man as he transforms his surroundings. Yet by paradox Fuller is much more at home in contemporary America than is Mumford.

Nevertheless these two Americans share many of the highest values of the national heritage. Fuller, after all, is the grandnephew of Margaret Fuller, the great feminist and friend of Emerson who has been described as the high priestess of transcendentalism; Mumford is the chief heir of Emerson in American moral philosophy. Each of them thus retains deep affinities with an older, more spacious, less mechanized, relatively unspoiled, and in part wild and almost unpeopled America in which individuals counted for more than organizations, which industrial technology has now changed forever. Yet each carries on the American intellectual adventure today, headed for different destinations but really voyaging outward to the world at large, following the great circle course of nonconformity, self-reliance, and transcendental awareness charted by Emerson and Thoreau.

But here any resemblance, either in philosophy or its practical applications, ceases abruptly. Precisely because of the breadth and scale of their synoptic outlooks—which Fuller, characteristically borrowing a term from astronomy, would call "sweep-outs" but which Mumford would consider valueless if not complemented by a searching "sweep-in" to the psychic needs of individual men—the two are irreconcilably opposed in their interpretations of man and his place in the universe, just as they are in total disagreement concerning the role of science and technics in civilized life. The profound differences between their concepts of planning and architecture are part of these larger philosophical differences, and cannot be fully understood except in the broader context of their total thought.

Neither man has dealt at length with the other in any of his voluminous writings, but everything to which Mumford would object in Fuller's approach is castigated in a powerful chapter in *The Conduct of Life* entitled "The Fallacy of Systems," which Mumford has called "a key to my whole life and thought." Since the seventeenth century, he asserts, "we have been living in an age of system-makers, and what is even worse, system-appliers," the Procrustean mutilators of life who lop off its essential irregularities to fit "a single set of consistent principles and ideal ends." But life "cannot be reduced to a system: the best wisdom, when so reduced to a set of insistent notes, becomes a cacophony: indeed, the more stubbornly one adheres to a system, the more violence one does to life."

Against "system-mongers" who seek "to align a whole community according to some limiting principle," Mumford opposes his "philosophy of the open synthesis," which he also names "the affirmation of organic life," in which the valid features of one system or another may be invoked by turns "to do justice to life's endlessly varied needs and occasions." This would be above all a life of "balance," maintained in "dynamic equilibrium," with an open attitude toward change. "Those who understand the nature of life itself will not . . . see reality in terms of change alone"; and "neither will they, like many Greek and Hindu philosophers, regard flux and movement and time as unreal or illusory and seek truth only in the unchangeable." History offers proof of this lesson: "Actual historic institutions, fortunately, have been modified by anomalies, discrepancies, contradictions, compromises: the older they are, the richer this organic compost." And yet, "all these varied nutrients that remain in the social soil are viewed with high scorn by the believer in systems."

Among the many systems, including capitalism and Marxian communism, that he enumerates with such distrust, Mumford has taken care to include utilitarianism, with its crass mechanistic notions of "efficiency." Although the social crimes committed by industrial civilization in the name of utility and efficiency should by now have shorn us of any simplistic mechanical "idea of progress," nineteenth-century utilitarianism happens to be alive and well in the twentieth, under the guises of technocracy and scientism; and if Mumford, in the rare passages where he mentions Fuller, had not considered him merely as an ingenious housing prefabricator with a "Jules Verne—Buck Rogers" side to him, but had instead analyzed Fuller's thought, he might well have consigned him to one of the lower circles of an organic philosophical hell among the Benthamites.

Fuller cannot be categorized quite so neatly, however, for he is an authentic original. Like Thorstein Veblen, he is much more than a rudimentary technocrat who has greater confidence in engineers and machinery than in politicians and the present price-system. But certainly, as far as faith in the system of doing the most with the least is concerned, he is a true believer. He is also an indefatigable system-applier who seizes any opportunity to find fresh exemplifications of his general principles, with the result that the earth's surface is now dotted with some five thousand geodesic domes of greatly varying dimensions and materials, serving innumerable purposes in climates as diverse as the Arctic DEW-line and steamy Louisiana, but all based on the same fundamental structural concept. As a system-monger he has long been incorrigible, and will cheerfully spend hours selling his ideas to anyone who will listen, from Tahiti to M.I.T., be they kindergarteners or U.S. Marines.

As physical systems go, Fuller's is remarkably complete, and yet, like his domes, structurally lean. For all his prodigality of language, replete with arcane Latinisms of his own devising that put a substantial obstacle between him and his audience, Fuller's system rests on only a few grand concepts (from which he draws a wealth of inferences), just as his architecture is limited to a few basic forms.

Except for some baldly technocratic borrowings, such as the contention that we had better give up politics altogether and let technology order the world, his philosophy is derived almost entirely from post-Newtonian science and is as near to being an *absolute* system, based on "universal laws," as any yet postulated by a modern environmental theorist. Since he began formulating universalist concepts in the twenties, taking the whole world—including the oceans and the atmosphere—as an integral field of development, Fuller's energies, according to his biographer and friend Robert W. Marks, have been "centered in a single drive: to promote the *total use of total technology for total population*, 'at the maximum feasible rate of acceleration.'"

But this unremitting exploitation of technology is inseparable from science and is based on cosmic harmonies. Fuller's structures are all based ultimately on the triangle, or combinations of triangles, which he claims is the basic building unit of the universe, lending itself naturally to incorporation in spheres, on macrocosmic and microcosmic scale. If this sounds like latter-day music of the spheres, reminiscent of the trinitarian metaphysics of the Middle Ages that found the triangle both universally symbolic and exceptionally stable as a structural component of vast cathedrals—or again, reminiscent of the theological associations of the purely proportioned domes of the Renaissance humanists—Fuller replies that such grand, simple forms express fundamental truths of nature, which he asserts is never arbitrarily complex. On the contrary, nature is accurately described by equations of the purity of $E = MC^2$. If Fuller is right, we are about to learn that chaos is nonexistent: "all that was chaotic was in man's illiterate and bewildered imagination and fearful ignorance."

But certain things will remain unpredictable, at least for the time being, due to the phenomenon that Fuller has dubbed "synergy," which means that "the behavior of a whole system" is "unpredicted by the behavior of any of its components—or sub-assemblies of its components." An example of a totality adding up to much more than the sum of its parts is a metallic alloy such as chrome-nickel steel, which, as Fuller gleefully points out, is far stronger than the sum of the strength of its separate ingredients.

A rather more overwhelming example of synergetics, to be sure, is the release of atomic energy. Fuller in one of his great insights foresaw in the 1930's that, thanks to a new and unlimited source of energy, a revolutionary increase in the world's productive capacity was at hand. With somewhat less insight, he assumed a corresponding improvement in human welfare.

What followed of course was the Bomb. Fuller was undaunted, and not really surprised, for he had long been aware that the most far-reaching technological innovations of modern times, for better and worse, have often originated in weapons research and development. Hence his unashamed association with the Pentagon—to Mumford the pit of conscienceless scientific and technological perversion—which has led to the large-scale delivery of his domes, sometimes by helicopter, to military establishments in many parts of the world.

Meanwhile, as the Cold War continued, nuclear power stations—soon to be combined with the saline-water conversion facilities indispensable to Fuller's long-standing scheme to occupy the watery three-quarters of the globe—appeared in many countries. In 1958 Fuller told Nehru: "Science has hooked up the everyday economic plumbing to the cosmic reservoir." In other words, wealth is now nondepletable. It is really "universal energy," and therefore classical economics, like classical geometry, may be totally discarded.

Thus Fuller's search for a general solution to terrestrial problems has gone into celestial orbit. Unhesitatingly he has placed himself in the forefront of the effort to get out into space, in order to tap inexhaustible cosmic resources; and he is a consultant to the team of scientists and technicians who are designing the first moon colony. What Fuller hopes to bring back from the moon, for the good of mankind, is the first "little black box" containing a truly miniaturized mechanical system including tiny energy fuel-cells and a tiny waste-disposal system, which will revolutionize household management everywhere in the world—at a time when there is not even everyday plumbing in the hovels of India. Analogous reasoning has permitted him to take some comfort from the introduction of giant helicopters in the Vietnam war, a war that in other respects causes him pain but nothing like the outraged sense of horror and anguish that has made Mumford one of the foremost spokesmen for peace and disarmament in the United States. If these powerful aircraft were released for nonwar duty, notes Fuller (who as early as 1927 conceived a light-weight, ten-story apartment house of "wire-wheel" structure to be delivered by zeppelin, which first would drop a bomb to excavate a foundation crater), they could deposit still larger geodesic domes any place on earth, providing huge instant shelters for Eskimos or Zulus.

With deepening despair Mumford regards such "power-centered" infatuation with technics as pathological folly. It is not that Fuller's schemes, which Mumford once might have dismissed as science fiction, are unrealizable. It is that they are all too swiftly feasible in an age of unparalleled *technical* capabilities that are beyond effective human control; and not surprisingly Mumford likens the Space Age, or Nuclear Age (he bitterly notes that it is not called the Age of Man), to an automobile without driver, steering wheel, or brakes, but crammed with demoralized passengers hurtling full speed toward doom. To those who call him Jeremiah, Mumford replies that it was Jeremiah whose prophecies "so fatally came true." As for Fuller himself, if Mumford would trouble to single him out as a prototypical antagonist, he might describe him, as he once did Le Corbusier, as "deeply in harmony with the negative tendencies of our times."

For Fuller is probably America's most vociferous and energetic advocate of what Mumford denounces as "the myth of the machine"; the notion that technics, rather than man himself, is the central component of culture. By now this technological interpretation of human progress has a rather long history, dating from the early industrial period when Carlyle and others first called man a "tool-using animal." On the contrary, Mumford replies, man from his beginnings has been "preeminently a mind-making, self-mastering, and self-designating animal." Therefore, it is not the mindless machine (which, according to Fuller, works "more reliably than the limited sensory departments of the human mechanism"), but man in all the richness and complexity of his "symbolic" activities who must be the true center. Not technical advancement, but human development, which imparts *significance* to life, is Mumford's measure of social good; and he warns that the chief moral problem in every part of the world, regardless of "prevailing ideologies or ideal goals," is to clarify the tragically blurred distinction between good and bad.

Thus, in book after increasingly formidable book, from *The Story of Utopias* in 1922 to the early masterpiece *Technics and Civilization* of 1934 and last year's *The Myth of the Machine*, a great autumnal work of which only the first volume has appeared, Mumford has celebrated the "mindfulness" of man, whose supreme discovery—in all the history of invention—has not been tools, but his human self. Man's most superb technics has not consisted of material technology at all, but of his human ability to speak and to dream, to laugh and to weep, to sing and to love. Hence human development is scarcely confined to work, even though Mumford places considerable value on "joyful common toil" (which should not be mistaken for nostalgic agrarianism). The sum of man's rational and irrational activities includes "ritual, art, poesy, drama, music, dance, philosophy, science, myth, religion"—the list could be extended endlessly—which are "all as essential to man as his daily bread." The universe appears very different, Mumford remarks, once "the light of human consciousness, rather than mass and energy," is perceived to be the central fact of existence.

Then organic *growth*, rather than "inordinate power and productivity" and "purposeless materialism," emerges as the principal goal of a truly civilized society.

Mumford has never rejected the machine per se. He abhors only the "overgrowth" of technics at the expense of human needs and aspirations—like the overgrowth of inhumane and inefficient metropolises—because it has thrown modern civilization into perilous imbalance that cannot be righted by simplistic injections of more powerful technologies operating at higher and higher speeds.

Mumford is a spiritual Luddite who would break, and certainly cease manufacturing, many machines, especially the sinister engines that, governed by "invisible" electronic controls can now virtually declare war by themselves. Even if the automated missiles do not put an end to human existence, other seemingly less menacing machines may wreck all that to Mumford makes life worth living, condemning men to "mandatory consumption" of the limitless array of products—not at all necessarily "goods"—poured forth from the technological cornucopia, and at the same time manipulating man socially and politically. If Fuller cheerfully suggests that the communications revolution has made it feasible for the entire population to cast its vote on any issue every day, Mumford asks: What are the issues, who is to formulate them, and how much will elections affect fundamental public policy?

For Mumford is more profoundly aware than Fuller that the impact of the mathematical and physical sciences upon technology in the past century, which has seen the rise of psychotic totalitarianism on every continent, has indeed caused "a radical transformation in the entire human environment." He further realizes, as Fuller in all innocence does not, that this great transformation—which has equipped man with so many marvelous "mechanical extensions"—also transforms man psychically.

"Alterations in the human personality" are produced, Mumford demonstrates, as indeed they have been throughout history whenever man has decisively changed the face of the earth. But the scale of modern technics is unprecedented and represents the most sweeping transformation of man since the Pyramid age. The age of "megatechnics" has begun, and it is perfectly capable of creating the very environment that a "dominant minority" (with which Mumford would surely associate Fuller) is striving to establish with the assistance of a "tentacular bureaucracy": "a uniform, all-enveloping, superplanetary structure, designed for automatic operation" that, Mumford fears, will reduce man to "a passive, purposeless, machine-conditioned animal." In the last third of the twentieth century Mumford, perhaps the last in the apostolic succession of the great Anglo-American environmental humanists—Ruskin, Morris, and Mumford's own master, Patrick Geddes—despairs of the now almost lost chance to build a truly "bio-technic order," based on human scale and human needs and employing technology only for the fulfillment of man. Thus he awaits the appearance of "Post-Historic Man"—the phrase is the title of Roderick Seidenberg's brilliant book—congealed in the icy fixity of a totally organized and utterly depersonalized technological age of "megamachines."

Yet this is a lesson Mumford has learned from his study of remote antiquity. Fuller's fantastic and exhilarating excursions into history have resembled raids, like those of the intrepid seafarers he so admires who adventured in "the outlaw area" on the high seas in the teeth of raw nature, where no bland social institutions hampered their expansion of human power as they took what they wanted and let the rest go. Mumford, on the other hand, is the greatest historian of urban civilization from its origins in the neolithic village to the "insensate" industrial metropolis; and in his grand historical synthesis he has traced the coexistence of two different technics: "one 'democratic' and dispersed," which found its highest expression in the authentic "polytechnics" of the Middle Ages; "the other totalitarian and centralized," whose supreme monument was the Pyramid of Cheops at Giza, the largest of man-made tetrahedral structures—Fuller's ideal assemblage of triangles. From the beginning, then, Mumford's "positive" and "negative," life-living and death-dealing technologies existed side by side, although—in some cultures—one or the other might be overwhelmingly dominant. Mumford also points out that unchanging, "static" technology, as in the case of bowls and cisterns, dams and reservoirs, and other "containers" (even the city is described as a "container of power"), is fully as important as "dynamic," incessantly changing technology. To this Fuller might reply as he did in one of his poems, "Change is normal/thank you Albert!"

In considering the Great Pyramid, which even by modern standards was built with extraordinary speed and precision, without heavy lifting machinery, Mumford had one of his great insights into the nature of technics. For the Pyramid was built by a "megamachine," which until Mumford's discovery had been "invisible" to historians because its thousands of interacting components were human. Society itself had been mechanized into an enormous machine capable of swift performance of enormous but essentially meaningless physical tasks; and there is an inevitable analogy with the largely "invisible" technologies in which we have invested so heavily today. What are the Egyptian pyramids, Mumford asks, "but the precise static equivalents of our own space rockets? Both devices for securing, at extravagant cost, a passage to Heaven for the favored few."

With unerring consistency Fuller has just completed Mumford's argument by designing a pyramid greater by far than any dreamed of by the Egyptians: his "Tetrahedral City," two hundred stories high and two miles long to a side, which will be a metropolis in itself. The vast structure, surely one of the most astounding visionary designs in the history of architecture, can either float on a moat in Japan, the earthquake-ridden, land-short country for which it was designed; or else it can be towed out on the ocean, freed at last from the chaos of history, with its nuclear power plant providing the energy to work its innumerable machines and its fresh-water distillery, and with supersonic airliners, giant ocean-going vessels, and electronic communications connecting it to the rest of the world and to outer space, where a Fuller-influenced settlement may by then be in operation on the moon. Here, as in the Great Pyramid, posthistory may begin.

Fuller's megastructure, by far the most ambitious project of its kind ever seriously proposed, weighing only a small fraction of the enormous tonnage of materials that would have been required, say, for Frank Lloyd Wright's "mile-high" skyscraper, is to Mumford nothing more than an "urban hive," better fit for social insects than for men. Rather, it might accommodate the superb robot that Fuller described, only partly tongue in cheek, in his answer to the question "What's a man?":

> Man?
> A self-balancing, 28-jointed adapter-base biped; an electro-chemical reduction-plant, integral with segregated stowages of special energy extracts in storage batteries, for subsequent actuation of thousands of hydraulic and pneumatic pumps, with motors attached; 62,000 miles of capillaries; millions of warning signal, railroad and conveyor systems; crushers and cranes . . . and a universally distributed telephone system needing no service for 70 years if well managed; the whole, extraordinarily complex mechanism guided with exquisite precision from a turret in which are located telescopic and microscopic self-registering and recording range finders, a spectroscope, *et cetera*. . . .

All that is omitted is the *purpose* for which this most beautiful mechanism exists; and not surprisingly, Mumford asks instead: What is the human brain? It serves:

> as a seat of government, a court of justice, a parliament, a marketplace, a police station, a telephone exchange, a temple, an art gallery, a library, a theatre, an observatory, a central filing system; and a computer: or, to reverse Aristotle, it is nothing less than the whole polis, writ small.

Only the telephone exchange, the complex communications system that both men have used with such courage, is mentioned in both quotations.

Yet the opposition of Mumford and Fuller is not quite that simple. If Mumford's generous philosophy theoretically accommodates incessant change while retaining the usufruct of the past, he himself has not accepted the largest part of the magnificent liberating innovations of the past generation. *The City in History* of 1961, in its recommendations for remaking the environment, did not differ in its essentials from *The Culture of Cities* of 1938. In an era of abundance Mumford called for the same program as in an era of scarcity; and although the program was handsome, proposing deconcentration and decentralization on the model of the remarkably prescient New York State regional plan of the late twenties, the British and Scandinavian new towns, and TVA, it took cognizance, for the most part, only of negative change during the interval between the two books.

Like the angry students at Berkeley, Mumford cried out, with reason, that our immediate need is peace, and that the human soul is not an IBM card. But the computer is inherently no more evil than the abacus, just as the jet plane is for no fundamental reason a less desirable form of transportation than the railroad. The value of machines depends upon the use to which they are put by men. In this respect history, on which Mumford has built one of the most complete philosophical syntheses of the modern age, may have escaped him. For the innumerable appliances and gadgets produced in the past thirty or forty years—which, together with birth-control pills, for instance, have opened an altogether new and almost incredibly more dignified life for women—have *not* thus far resulted in what Mumford calls "a dismally contracted life, lived for the most part confined to a car or to a television set."

Thirty million people, in this nation alone, remain impoverished, but the majority of Americans know that machines, besides bringing air pollution and devastation of the landscape, have brought them their new leisure, their release from age-old drudgery, their improved education, health, and cuisine, their ability to travel and to sail in their own boats, their symphony orchestras, galleries, and museums, their growing awareness of the necessity for privacy and solitude, their fresh awakening to the majesty of the wilderness, their educational television stations, and their millions of inexpensive paperback books, including Mumford's and Fuller's.

Fuller has ridden this wave of the future, remarking that "the old-fashioned square-shooter is today's square." If his puristic simplifications are oversimplifications, if his prefabricated houses and bathrooms are not as good as fine traditional dwellings as yet, if his domes and space-frames impose standardized solutions where a fully industrialized technology would provide limitless variety for personal choice (as it does in our machine-made clothing), Fuller has nevertheless been open to the future.

And the future remains open to us. To adventure truly upon the future as civilized beings we need new criteria for action: an unprecedented philosophy capable of solving unprecedented problems that neither would be linked indissolubly to the past (for it is conceivable that on positive grounds we may wish to relinquish much more of the past, which, after all, has landed us in the present predicament) nor would be stained with the philosophical sin of pride that Bertrand Russell called "cosmic impiety." The world, as Mumford knew from the start of his career, is a single complex entity, but it must be ready for change on a scale that few men, Fuller among them, have dared to contemplate. What each of them has done, really, has been to write philosophical poems celebrating a world that does not truly exist, and perhaps can never exist, even though the poems are true. Mumford is an epic poet, as grave, as moral, as grandly tragic, as John Milton; Fuller is a lyricist, and his bright, luminous structures had best be taken as lovely technological songs. Someday, from somewhere on the unified earth, a new poet may emerge to combine their gifts; but that supreme poet, as Santayana wrote at the end of his appreciation of Lucretius, Dante, and Goethe, is in limbo still.

INTRODUCTORY NOTE

Theodore Roszak wrote "Technocracy: Despotism of Beneficent Expertise" when student radicalism and the antiwar movement in America were strong. His attitude toward technology and science, and the society shaped by them, is representative of the antiquantification, antirationalism, and antielitism of the late 1960s. But Roszak's sentiments, even though shared by many Americans when he wrote, were contrary to a tide of opinion that had run strong for generations and that is expressed in many of the readings in this book.

Roszak directed his attack at the spirit of technocracy, which in his view was the essence of the engineering, managerial, and scientific mentality ruled by imperatives of efficiency, economy, and production rather than humanistic values and esthetic concern. He believed that American society was moving toward technocracy along many paths and that it was a massive historical movement—"the wave of the future."

Interestingly, Roszak looked back, not without nostalgia, to the engineers like Edison and the scientists like Galileo who flourished before scientists and engineers had "tasted power and corruption." In this instance, Roszak's judgment was shaped by myth, for as the articles in Parts III and IV show, scientists and engineers of the past associated themselves with the very attitudes that Roszak so bitterly condemns in their "technocratic" counterparts today. Buckminster Fuller and Simon Ramo, an engineer, whom Roszak believed representative of the malaise of the sixties, expressed attitudes similar to those of the scientists and engineers Robert Millikan, Arthur D. Little, Robert Thurston, and Thomas Ewbank decades earlier. Millikan, Little, and Ramo, though a generation apart, called for the expert user of scientific and technological methods to apply these to a range of societal problems extending far beyond the simply technical and purely scientific. Fuller, Thurston, and Ewbank, writing in different centuries, were enthusiastic at the prospect of a man-made world.

Theodore Roszak (1933–   ) is the editor of *The Dissenting*

Academy (New York, 1969), which has essays by leading radical scholars including Lynd on history and Chomsky on the responsibility of intellectuals. He is also the author of the widely read *The Making of a Counter Culture: Reflections on the Technocratic Society and Its Youthful Opposition* (New York, 1969) and *Where the Wasteland Ends: Politics and Transcendence in Postindustrial Society* (New York, 1972).

Technocracy: Despotism of Beneficent Expertise

THEODORE ROSZAK

Our society advances toward technocracy along many paths. The authors assembled here for review help us count the ways. Some—bravura technicians like Buckminster Fuller—crusade flamboyantly toward the regime of scientific expertise, flying banners in behalf of an *outré* social engineering whose goal is nothing short of "world planning." Others, like the members of the Pittsburgh Values Project, smooth the way more subtly by laboring to convert our capacities for evaluation and taste into a new behavioral technics accessible only to the academic specialist. Some, like the industrialist Simon Ramo, seek to clear the obstacles of traditional politics from our line of advance by entrusting our social problems to the objective competence of the systems analysts. Others, like Michael Reagan investigating the nearly total dependence of science on federal patronage, speed us along our way by celebrating the virtues of the alliance, while assuring us that any vices accruing can be easily adjusted by vigilant administration within the corridors of power. Even those like Victor Ferkiss who view the journey with a critical eye, contribute to our progress toward technocracy by conceding far more to the technician's world-view than their humanistic instincts should permit. The result is frail resistance in the face of a massive historical movement whose seriousness we underestimate at the peril of our human dignity.

No doubt there is room to disagree about the extent to which a technocratic totalitarianism has been developing in our society. I believe we have gone too far in this direction and that the technocracy, despite the stubborn survival in our midst of sundry flat-earthers and dilettante occultists (as well as a growing population of hip young swamis), looks more like the wave of the future for all industrialized nations, regardless of official ideology, than anything else now in sight. In any event, a lively

The Nation, 209 (September 1969), 181–188. Reprinted by permission.

concern for that prospect is the best touchstone for evaluating the burgeoning literature on technology and society. Is the writer aware of the technocratic possibility? Does he welcome or resist it? Is he aware of the force technocratic tendencies exert upon our politics and our total culture, of the costs and losses to which they can lead? Such are the questions I find myself persistently asking whenever I come upon still another diagnosis of our high industrial agonies.

Technocracy, Jean Meynaud says in his important anatomy of the beast, stems from "the pressure of technology on the political system." It is the demand "that politics be reduced to technics." The seed from which technocratic politics sprouts first blossoms in the field of industrial invention and organization, the obvious province of technical intelligence. But the high yield which man's engineering talents achieve there, bolstered by the technician's claim to scientific rationality, eventually encourages the same habits of mind to embrace social life as a whole. As Meynaud puts it:

> Founded on the advances of scientific thought . . . modern technique becomes, in its broadest sense, all those methods which allow man to make best use of existing resources to satisfy his material needs and common ideals. . . . Technique is not confined to a limited sector of society: it is society looked at from a certain perspective.

The perspective, that is, of technician and scientific intellectual.

Regarded in this light, as a product of both the scientific and industrial revolutions, technocracy assumes a formidable cultural momentum. Its legitimacy reaches deep into all that modern Western man has come to consider good, true and beautiful: affluence, empirical knowledge, and prodigious material power over his environment. As a social form, it is the irresistibly logical consequence of what science tells us reality is and of what technicians tell us can rationally be done to manipulate that reality.

The essays in Lynn White's *Machina Ex Deo* provide a keen examination of the larger cultural context out of which our science and technology evolve. His book—which includes the superb essay, "The Historical Roots of Our Ecological Crisis"—is a model of concerned and elegantly literate scholarship for the general reader. One could not ask for a more sensitive approach to the now submerged religious origins of industrialism: "the sources of our faith in science today and . . . the wellsprings of motivation that lead man to pursue science."

White is far more sanguine about the prospects of a "new humanism" arising among the engineers than I can be. As a medievalist, he tends to see scientists and technicians as still the free spirits and rugged individuals they were before they tasted power and the corruptions attending. The transition from "little science" to "big science" (to use Derek de Solla Price's terms) has taken a greater toll of the pristine ideals of scientist and engineer than White may realize. How much of the Christianity of the catacombs survived into the age of Innocent III? Perhaps as much of Galileo and Edison survive in the age of Edward Teller and C. P. Snow. Still, White is correct in identifying technology as "a prime spiritual achievement," an expression of mankind's tireless search for "a new order of plenty, of mobility, of personal freedom." It is only by drawing upon such vital desires that the technocrats could achieve the great moral and psychic power they hold over our allegiance. Technocracy, far from being only an opportunistic and cynical plot (and the worst social evils never are), is a proud, noetic citadel solidly based on all that scientific inquiry has accomplished after the dark millennia during which man was supposedly sunk in superstition, mythological guesswork and helpless bewilderment. The policies that emanate from the citadel are born of pure reason and generous intention; its custodians are those who qualify as disinterested experts in all aspects of life—or, at one highly important remove, those who control the services of such experts.

One senses how domineering the regime of experts has become when one recognizes the lengths to which contemporary radicalism must go in seeking to outflank its values and metaphysical assumptions. An eloquent example of such an effort is Gary Snyder's *Earth House Hold* which seeks to recall the ecological intelligence to be found in poetic and primitive life styles. For Snyder, who learned his social theory from Zen masters and redwood trees, our salvation depends upon recapturing the spirit of "the ancient shamanistic-yogic-socioeconomic view" in which there abide "the most archaic values on earth: . . . the fertility of the soil, the magic of animals, the power-vision in solitude, the terrifying initiation and rebirth, the love and ecstasy of the dance, the common work of the tribe."

The audience for Snyder's rhapsodic appeal is small, largely the young and the dropped out who cling on and make do, chanting mantras at the social margins; "the tribe" as Snyder calls them, referring to those like the sod brethren of the Berkeley People's Park whose delicate experiment in Arcadian communitarianism the Governor and University of California have determined to trounce out of existence. Unhappily, visionaries like Snyder are regarded as the heretics-in-residence of our pluralistic technocracy, and their words make light weight in the scales of power. Not that society is closed to visionary declarations. Visionary *technology*—of the World's Fair, science-fiction variety—easily wins spellbound regard; for the millions, it monopolizes the whole meaning of imaginative daring. They will starve their poor to finance Project Apollo; they will sacrifice their peace of mind to have supersonic airliners split the skies above their cities with the sound of progress; they wait with bated breath to have their gene pools programmed for the guaranteed production of talented and beautiful progeny. Those who speak the language of numbers and technologisms never lack for reverent if bewildered public attention, and the bolder the scientific impresarios become, the greater the awe and acquiescence. One hundred million Americans may attend church every Sunday, bringing with them superstitions and prejudices as old as the Stone Age. But when they exclaim rhetorically, "Will wonders never cease!" it is the works of the engineers and not the works of God they have in mind.

Buckminster Fuller surely ranks as dean of the Promethean technicians. Certainly few exert greater personal magnetism among the "think big" younger generation of architects and engineers. Among Fuller's great schemes: his project to house the human race in prefabricated, helicopter-transported geodesic domes (Fuller's most famous and most overpraised invention) and the birth-certificate-universal-credit-card designed to operate anywhere on earth by means of a frequency tuned to the holder's genetic code. Now, in Fuller's latest book, the earth is stripped of its metaphorical motherhood and becomes instead a spaceship in need of nothing so much as properly trained "world planners" in its cockpit. The technocratic world-view could not ask for a more serviceable central image—though it will certainly require a less idiosyncratic "operating manual" than Fuller has to offer.

Some of Fuller's notions—like his "synergetic mathematics"—may have vaguely mystical implications. But he is careful always to protect the respectability of his bizarre brainstorming by disguising it as a form of omni-competent, global engineering. Like Marshall McLuhan, he is a great bamboozler: a combination of Buck Rogers and Horatio Alger. His intellectual repertory rarely includes a reference to artist, philosopher, poet or prophet. This, along with his anti-academic autodidacticism, makes him decidedly the technocrat Old Style. Yet he manages to lay ham-handedly on the line most of the tacit assumptions of the technocratic mentality: that it was only when man became a machine maker that he "began for the first time to really employ his intellect in the most important way"; that invention and efficient organization, embracing as they do the whole meaning of Reason and Progress, necessarily improve with increase; and that for every human problem there is a technical solution—sometimes a dazzlingly simple one. Thus: "You may . . . ask me how we are going to resolve the ever-acceleratingly dangerous impasse of world-opposed politicians and ideological dogmas. I answer, it will be resolved by the computer . . . all politicians can and will yield enthusiastically to the computer's safe flight-controlling capabilities in bringing all of humanity in for a happy landing."

To be sure, "General Systems Theory," the form of technical expertise Fuller recommends for the better piloting of "spaceship Earth," is intended to be ambitiously broad-gauged: the very antithesis of the myopic specialization which is Fuller's bête noire. The prospectus sounds intriguing, but the social realities behind GST are bound to be unappealing to those who have sentimental attachments to participative and communitarian politics. This form of elitist brains-trusting—an outgrowth of wartime operations research—is apt to be the essential component of the technocracy's *machine à gouverner*.

Those who desire a brief, ebullient account of the systems technique may consult Simon Ramo's *Cure for Chaos*. Ramo, vice chairman of Thompson-Ramo-Wooldridge, father of the ICBM, and, as the book jacket tells us, "frequently consulted by government leaders," is, like Fuller, a self-made technician-entrepreneur. The beneficiary and booster of military-industrial boondoggling, his vision of life is colored accordingly. The archetypal social problem for Ramo is patterned upon the gadgeteering perplexities of the ballistic missile system, where everything, including the megadeaths, yields to simple numbers. Ramo readily admits that, in matters involving "the human component," the "systems team" must include "techno-political-econo-socio-experts." Yet systems engineering discriminates not at all between problems of air traffic control and education, the arms race and urban renewal. The problem-solving procedures are identical. One assembles the certified experts and trusts them to convert the project at hand into "specific well-described and often measurable performance requirements." The skill of experts, Ramo tells us, lies in "putting quantitative measures on everything—very often cost and time measures."

Thus, in quantification the systems engineer has found what the alchemists of old sought in vain: the universal solvent. It is little wonder that the fiercely mathematical passion of such expertise finds no trouble in sliding from discussions of thermonuclear weaponry to hospital design in successive paragraphs, or in conceiving of education as something whose value is calibrated in lifetime income differentials and whose substance can be captured at "electronicized desks," where vigilant computers pace their attentive students through true-or-false and fill-in-the-blank testing.

After Ramo's and Fuller's intoxicated high touting of the systems approach and computerized instruction, Anthony Oettinger's *Run, Computer, Run* is sobering reading. Oettinger wisely works from the assumption that the Himalayan dilemmas of American education are through and through political and philosophical. This means that "systems analysts trained to think unthinkable, apocalyptic thoughts in the style of Herman Kahn, or to calculate the performance/weight tradeoffs for missiles, are ill-prepared to deal with more than the form of the educational system."

> The present tools of formal systems analysis work best on well-defined, simple, concrete models involving quantifiable con-

> cepts, measurable data, and, above all, thoroughly understood theoretical structures which adequately reflect reality.

Since almost nothing about education, beyond perhaps the plumbing of school buildings, yields such convenient simples, Oettinger spends little time on the latest instructional gimmickry and systems techniques. Rather, he devotes himself to debunking the many "highly visible quickie" solutions currently on the scene that offer only "the illusion of progress." One by one, Oettinger undermines the widespread assumptions that teaching machines save money, save time, save manpower, and "individualize" instruction. It is a commendable display of candor from one who, as director of Project Tact ("Technological Aids to Creative Thought": Harvard based, Defense Department financed), would seem to have every interest in extolling the myths here marked out for destruction. Oettinger's enormously intelligent book is a commendable product of Harvard's Program on Technology and Society. Indeed, the program would seem to have its future course well surveyed if it were to extend Oettinger's critique to the relations between technology and society generally.

Clearly the job needs doing. It is wholly disheartening to review the evidence Oettinger presents of our society's scientistic gullibility and gadget mania—in this case from the U.S. Office of Education down to the smallest local school board—and of the cynical flummoxing perpetrated by industry and eager-beaver systems analysts. In education as in all other areas of our life, we are indeed patsies for what Oettinger calls "the seductive power of absurdity in full formalized attire." And, sad to say, as sharp a critic as Oettinger himself can have his blind spots in this respect. Even he grants that systems analysis, for all its limitations in the field of education, *is* properly at home in the Defense Department. Quoting Charles Hitch (now President of the University of California, via RAND), Oettinger agrees that it "is easier to program and to analyze [military strategy] quantitatively than many areas of civilian government. For example, it is certainly easier than the foreign affairs area."

That is a fatal, but revealing admission. For after what fashion has Defense Department policy become more adaptable to systems analysis? By subordinating to its oppressively quantitative and value-neuter demands every aspect of life that gets in the way (including "the foreign affairs area"). In brief, systems analysis has "worked" for our society's infinitely expansive conception of military necessity precisely as it purports to "work" for our educational problems: by flattening every non-quantitative consideration beneath a steam roller of dehumanized logicality.

It is the great claim of the systems approach that it embraces *all* facets of every problem. It purports to blanket the whole with coordinated competence, leaving nothing to intuition or amateurish improvisation: meaning, it leaves nothing to the layman. This claim to comprehensiveness, however, is fundamentally false. Systems methodology by its very essence stops short of taking into critical consideration the whole of society, economy, environment, or the human condition. Such wholes are what Ramo calls "too big a system." He warns us:

> Surround the problem too broadly, try too hard to be absolutely complete, and you will not only get nowhere in the solution of the problem, but you will be doing a terribly poor job of systems engineering.

In what sense are such systems "too big"? Why, too big to be handled in the specialist's value-neuter framework. To think on such a scale is to aspire beyond amoral measurements and partial adjustments toward philosophical comprehension and social criticism, and doubtless toward many embarrassing questions about the likes of Simon Ramo, his fortune and influence in contemporary America.

It is revealing of the technocratic temperament that Ramo should finish his treatise by projecting the "golden age" that will ensue "once most people are wedded to creative logic and objectivity to get solutions to society's problems." Logic and objectivity: translated (with a deal of legerdemain) from mathematics and physics into the professional study of society, are science's gift to the apologetics of technocracy. One does not moralize about microbes and quasars; *ergo* logic dictates that the objective mind does not moralize about Thompson-Ramo-Wooldridge, thermonuclear deterrence, or the uses of the multiversity. Where the "human factor" is concerned, the systems team may, as Ramo puts it, have to "tap preferences, judge needs, present possibilities, and evaluate alternatives." But it is the experts who will do the tapping, judging, presenting and evaluating: the "chaos" is theirs to cure. And the experts work neither for free nor in a vacuum. As Jean Meynaud makes clear:

> For technocracy to cease being essentially conservative, technicians themselves would have to become inspired with the will to change. . . . But it is not rare for the technician to consider that the social system (one which manifests itself with continuity) is necessary to the realization of his designs. After all, the supporters of Saint-Simon were not lacking in human generosity, but they finally gave the best of themselves to the big industrial undertakings and commercial banks.

For his own France, Meynaud is able to trace the roots of this conservatism to the family and educational background of the highly cliquish Gaullist polytechnicians. Though they are more diffuse in social origin, the regents of the American New Atlantis display the same combination of "technical boldness and social conservatism." What passes for objectivity among those who staff our military-industrial think tanks is an understandable reticence respecting the world-view and social system that justifies and rewards expertise. Nothing could be more quantifiable than the gold in Ramo's "golden age": it measures out in dollars and cents, status and prestige.

This same practiced evasion of all philosophically challenging questions characterizes Michael Reagan's examination of science under the patronage of the government, a study that never once confronts the overarching issue: the political form that must inevitably follow from the progressive assimilation of expertise by forces commanding the bulk of society's cash and privilege. Indeed, Reagan would press the process of assimilation further still. He calls for the funding of a National Foundation for the Social Sciences (on the order of the National Science Foundation) as the best way for social scientists to make "their developmental needs known to government through a regularized, internal, differentiated organizational position" and for them "to be in on the choice of problems and of strategies for dealing with them."

What news it would be to Plato and Marx, to Veblen and C. Wright Mills to learn that students of man and society must make their "needs known to government" and must wait upon official moneys and favor to choose their problems and strategies. But clearly this is a rather different breed of student, one who approaches the Prince with the same power-political promise that has served as sweet bait on the atomic scientist's hook.

Reagan's conception of political science is contained entirely within the corridors of power, where everything that is, is —with minor adjustments—right. Inevitably, then, we reach the cheerful conclusion that the scientists, the multiversities and the government enjoy a "fruitful relationship" bearing "all the earmarks of permanence"; and if there are any problems, they can be rectified by some reshuffling of agencies and funding priorities. Reagan's first priority would go to "social objectives which are defined as most urgent politically." He does recognize, though without alarm, that this process of "definition" is often carried out by "vested interests in government agencies and outside groups." The remedy for this vice, however, is another echelon of supervisory expertise . . . a Technology Assessment Board perhaps . . . or an amalgamation of the Office of Science and Technology with the President's Science Advisory Committee . . . or why not an expansion of JCST . . . or of COSPUP . . . or of OSRD . . . etc., etc.? Reagan's study is a fine reference guide to the C. P. Snow world of the science bureaucracy. But with respect to all the great questions of justice and democracy and intellectual conscience—the issues that drove Norbert Wiener out of the "science factories," that sent Einstein heartsick to his grave, and that have stirred student rebellions across the land—Reagan is totally "objective." His suggestions for reform are eminently practical. He goes no further than to play committeemanship with the expert-administrators.

It would be wrong, however, to suggest that the technocratic mentality at its best lacks concern for values. On the contrary, the professional study of values is, along with technological prognostication ("*futuribles*," as the French call it), a hot and highly sophisticated specialty. A volume like *Values and the Future* serves as a twin introduction, authoritative if ponderous, to both the new science of ethics and the "futures industry."

The anthology grows out of a high-powered conference held at the University of Pittsburgh during 1965–66 under grants from IBM and the Carnegie Corporation. The result is academically suave, but hardly encouraging. With the exception of a few of the less pretentious essays such as those by John Kenneth Galbraith and Kenneth Boulding, the anthology deals in the learned superficialities which are the hallmark of technocratic culture: the systematic circumvention of emotional engagement and personal commitment. The essays are filled with baroque methodologies, much statistical virtuosity and esoteric computer games; and these last, as described, achieve levels of absurdity rivaling Ionesco. (Thus, in one of the games, the conferees, in the pursuit of computable statistics, take it upon themselves to "simulate" teen-agers, housewives, cultural elite and "persons in lowest income decide" as of the year 2000. "And *that* is science, my friend," said Dr. Spalanzani.)

Professionally, it is all quite elegant. But what is it, at last, that technical expertise does with values when it gets hold of them? It treats them like alien organisms under a microscope. It surveys them by way of foolish questionnaires no serious person would agree to complete. (Imagine Tolstoy or St. Paul being asked to give their "relative preferences" and "numerical evaluations" regarding "personality control drugs" and "household robots.") It quantifies them, by way of "weightings" and "scalings" that are the mathematical analogue of a Rube Goldberg contraption. And finally, it predicts their future course, by way of facile computerized techniques—in this case, their course under the impact of technological forces that are taken to be as inexorable as the will of God.

Not very remarkably, "the findings" that come of such procedures are always upbeat. "The most striking feature of the questionnaire responses is their clear optimism," Nicholas Rescher reports on one of his surveys. But who were the respondents? Fifty-eight experts from IBM, RAND, NSF, and the IBM-financed Harvard Program on Technology and Society. (Are there no laws against such intellectual incest?) Again, to quote Meynaud, "the future is influenced by the picture which men themselves draw of it. . . . One of the trends of 'futurism' is a tendency to exalt the virtues of a technical civilization and the merits of technicians. . . ."

In the introduction to *Values and the Future,* Alvin Toffler declares that "we need systematic surveys of the value systems held by important American subcultures, by different professions, age groups and socioeconomic groupings." It is surely a call for a million years of "extensive—and expensive—empirical research." But, Toffler feels, only this will give the specialists what they need as they set about "drawing up their reports in forms that can be taken into account in cost-benefit appraisals." Ideally, he anticipates a new profession of "Value-Impact Forecasters" who will be "armed with scientific tools" and "located at the hot center of decision-making" as "part of every corporation, research laboratory, government agency and foundation whose output includes technological innovation."

Surely it is the perfection of technocracy when values become statistical grist for the research mill. Once such a great project is launched, shall we not then be assured that all the goods and evils of policy and decision have been taken competently into account—with logic and objectivity—by the court wizards? And what will there be left for the layman to worry his amateurish head about?

How does one drive it home to such academic cold fish that their project, serviceable as it is for purposes of technocratic consolidation, is misconceived *ab initio*? As Socrates knew 2,500 years ago, to enter the agora simply to *survey* the so-called values of a befuddled public is the betrayal of philosophy. The values of men are not to be measured or predicted but to be honestly debated, affirmed and deeply lived, so that we may educate one another by mutual example. It is *this* that we owe one another as fellow citizens. But I doubt that this distinction between the academic and the irreducibly existential would prove persuasive. Expertise, being committed to that self-congratulatory form of alienation called "objectivity," makes no allowance for the person. It discounts the experiential deeps and attends to the behavioral surface. What chance would there have been to convince Dr. Kinsey that he had it all wrong? That love is not to be measured but to be fallen into, made, enjoyed and suffered? "But," the scientific mind insists, "there is *also* matter here (and where is there not?) in need of professional research. For only by professional research do we come to *know* anything."

And that is the box in which the technocrats have us trapped: their conviction and ours that only the scientific mind—objective, logical, quantitative—has access to reliable knowledge. What other remedy can there be, then, for the ills of mankind in a scientific culture but still more science, still more technology? Was it not scientist and technician who constructed our industrial cornucopia in the first place? Who else can understand its mysteries and maintain its flowing abundance? And maintaining industrial abundance is what everybody's politics is about in our time—Left, Right and Center. It is precisely this epistemological monopoly that legitimizes the shameless self-selection and internal circulation of technical elites within our military-paramilitary-financial-industrial-multiversity-think-tank-foundation-Presidential-advisory complex.

Here is the point that I believe Victor Ferkiss overlooks when he writes off the possibilities of technocratic totalitarianism. Granted, he is correct in concluding that the American version of the technological society really overlays and serves old-fashioned, bourgeois profiteering. (Technocracy has no difficulty absorbing any social system, provided the system is committed to an expanding economy.) Granted too, he is correct in concluding that the effort to create a totally efficient technocracy—a Huxleian Brave New World—is more apt to produce a "clumsy monster" constantly on the brink of chaos. But he is wrong in deciding that these facts make "the notion of a scientific elite a myth," rather than a growing reality. For what does the citizen in a scientized culture have by way of knowledge or value to hold against those who are the experts or those who own the experts? Values, as we now learn, are becoming the province of ethical technicians equipped with inscrutable methodologies and expensive computers. Knowledge—the product of logic and objective research—is similarly the property of the experts. True, we may still resist the experts and their employers. But by appeal to *what* does our resistance lay claim to the sanction of reason, sanity or progress? What is there the citizen can claim to know that an expert somewhere cannot prove he knows better?

I believe that Ferkiss underestimates this mystique of expertise because he himself has fallen beneath its spell. Even he, an admirably humanistic critic, believes that the thrust of science and technique "by giving man almost infinite power to change his world and to change himself" has carried us to the threshold of nothing less than an "evolutionary breakthrough," to which all culture must adapt if we are to produce the mature "technological man." Once concede that this is the proper relationship of science-technology to culture (the former disposes, the latter scurries to adjust) and our political life is delivered into the hands of the technical experts and their patrons. All complies to their world-view and there are no alternative realities. No doubt the technicians *will* proceed to build us a "clumsy monster" of a social system, a push-button bedlam with music by Muzak; but they will also convince the citizenry—by logic and objectivity—that it is the only rational system. Jean Meynaud's analysis is sharper than that of Ferkiss, and much gloomier.

> We might wonder whether the determination to obtain maximal efficiency, combined with greater prosperity in larger sectors of society, is not likely to throw into gear a movement which will lead to progressive monopoly of power by "competence." Supposing that such an evolution occurs, the final end of the scientific society might perhaps be democratic, but its functioning would no longer be so. In the extreme case, a scientific technocracy . . . would replace traditional political machinery, conserving the essential nature of the established order.

One must face the fact that such a despotism of beneficent expertise, guaranteeing both the bread of cybernated affluence and the circuses of moon shots, would, at this point, be far from unwelcome to most of the Americans it was willing to integrate. Perhaps only a handful of wild and woolly dropouts and stubborn black nationalists would be inclined to resist its invitation to the fat and dependent life. Certainly Meynaud's proposed safeguard—an unspecified strengthening of elected authorities and their institutions wholly within the going industrial system—looks like a very weak reed. This is because it ignores the psychic and metaphysical depths to which the technocracy reaches in achieving acquiescence and in this way undercutting democratic institutions.

Erich Fromm does greater justice to this dimension of the problem in his *Revolution of Hope*, an inspirational tract that provides a sensitive, if primer-simple assessment of our technological dehumanization. With a shrewd eye for the pathological, he tells that

> the main problem . . . is not whether . . . a computer-man can be constructed; it is rather why the idea is becoming so popular

in a historical period when nothing seems to be more important than to transform the existing man into a more rational, harmonious, and peace-loving being. One cannot help being suspicious that often the attraction of the computer-man idea is the expression of a flight from life and from humane experience into the mechanical and purely cerebral.

The major weakness of Fromm's treatise lies in the author's eagerness to dish up a rather thin potpourri of specific reforms (though there are several good ideas here) and to launch a new political movement. I doubt that the changes Fromm desires should, or will have the conventional organizational characteristics he suggests: a national committee of notables ("The Voice of American Conscience") reared on a base of local subcommittees. The "psychospiritual renewal" he calls for must perforce be more subtle and spontaneous, and surely less concerned than Fromm is with avoiding "the breakdown of the industrial machine." It will have to get at the millions as insidiously as it has apparently gotten at Fromm himself, whose latest writings begin to reveal a pronounced compromise with his former secularized, Socialist humanism, and a growing disposition to take seriously religious forms of community and personal transcendence.

Fromm does not, to be sure, move far enough in this direction to match the sensuous, primitivist mysticisms of Gary Snyder: "the timeless path of love and wisdom, in affectionate company with the sky, wind, clouds, trees, water, animals and grasses." And his hesitancy is understandable, since to cut so deep seeking the roots of the technocracy means hazarding one's intellectual respectability. But I think it has come to that desperate a pass. Caliban, having acquired the higher mathematics and devised him a clever machinery, hastens to counterfeit his master's magic. We may then need more than a touch of Blake's brave perversity

To cast off Bacon, Locke & Newton
from Albion's covering,
To take off his filthy garments & clothe
him with Imagination.

INTRODUCTORY NOTE

The recent and sharp change in American attitudes was clearly expressed in a concern for technology's adverse effect upon the natural environment. There has been no more effective spokesman for this opinion than Barry Commoner. His conviction that nature and its ecological systems must be saved is no less passionate than the nineteenth-century American's determination that the technological transformation had to take place. In a dramatic turnabout, Commoner was trying to preserve what his ancestors had enthusiastically subjugated and even eliminated.

Commoner realized—as did his contemporaries—that Americans had created an ugly and even hostile environment in many parts of their world, but he went beyond this commonplace to point out that they had not even achieved their desired independence from nature in these man-made places. Americans had believed they could substitute technology for nature—that synthetic materials and atomic energy, for instance, would free man from the uncertainties of weather and the limitations of natural resources. As Commoner pointed out, however, America's new materials and energy sources were still very much a part of the biosphere, the larger world that includes nature and technology. And technology can disrupt the balances of the biosphere and as a result pollute water, air, and food, some of its natural components for which man has not found adequate substitutes.

In stark contrast to nineteenth-century Americans who called for more machines, processes, and the sciences to support them, Commoner wanted a new emphasis on biology and sciences related to natural systems. Such a shift would help man to harmonize nature and technology rather than to attempt to conquer and subvert nature with technology. He also called for the man-made world to be designed with the quality of life rather than the quantity of production as the criterion.

Barry Commoner (1917–   ) as a research scientist and university professor has been especially concerned with the fundamental problems of the physiochemical basis of biological processes. A botanist at Washington University, St. Louis, after 1947, he became director there of the Center for Biology of Natural Systems in 1966. By 1970 he was well known as an informed and persuasive advocate of environmental policies. His books *Science and Survival* (1966) and *The Closing Circle* (1971) were analyses of the problems of technological society. Commoner convinced many Americans that the pollution of air, water, and earth by chemicals, especially the residuals of fertilizers, detergents, insecticides, and the profligate use of energy, has generated an environmental crisis of disastrous proportions.

Technology and the Natural Environment

BARRY COMMONER

The architect is the designer of places for human habitation. But the habitat of man is not merely buildings, roads, and cities. It is, rather, the earth's total skin of air, water, and soil, for it is that planetary system—the biosphere—which establishes the basic conditions that support the life of man. Like all living things, human beings can survive on the earth only so long as this environment is fit to support them. What the architect designs for the use of man must therefore fit into the design of the environment.

Until recently the environment has been largely taken for granted—that it will continue, as it always has, to support our life and our livelihood, providing the air that we breathe, the water that we drink, the food that we eat, and much of our industrial raw material. In the last few years, with a sudden shock, it has become apparent that modern technology is changing the environment—for the worse. The air that we now breathe in our cities can lead to respiratory disease, and lung cancer; surface waters are losing their natural capability for accommodating human wastes; environmentally-induced changes in food crops are causing disease in animals, and in some instances people; human activities may threaten—depending on the outcome of two contradictory effects—either to flood the cities of the world under water from the earth's molten ice-caps, or to induce a new ice age.

I should like to review briefly what has been learned from the mounting roster of environmental problems. The lesson, I believe, is simple and grim: The environment is being stressed to the point of collapse. I believe that we are approaching, in our time on this planet, a crisis which may destroy its suitability

---

The Architectural Forum, 130 (June 1969), 69–73.

as a place for human habitation. But I believe that we have also learned that the environmental crisis *can* be resolved if we accept a fundamental fact—that man is not designed to conquer nature, but to live in it.

The proliferation of human beings on the surface of this planet is proof of the remarkable suitability of the terrestrial environment as a place for human life. But, the fitness of the environment is not an immutable feature of the earth, having been developed by gradual changes in the nature of the planet's skin. Living things have themselves been crucial agents of these transformations, converting the earth's early rocks into soil, releasing oxygen from its water, transforming carbon dioxide into accumulated fossil fuels, modulating temperature, and tempering the rush of waters on the land. And, in the course of these transformations, the living things that populated the surface of the earth have, with the beautiful precision that is a mark of life, themselves become closely adapted to the environment they have helped to create. As a result, the environment in which we live is itself part of a vast web of life, and like everything associated with life, is internally complex, and stable, not in a static sense, but by virtue of the intricate play of internal interactions.

On a small scale, the dependence of environmental stability on the nice balance of multiple biological processes is self-evident. A hillside denuded of vegetation by fire, and thus lacking protection against the erosion of heavy rains previously afforded by the canopy of leaves and the mat of roots, can quickly shed its soil and lose its capability to support plants and harbor animals. And, on this scale, the threat of thoughtless human interventions is equally self-evident; we have long since learned that brutal lumbering or greedy exploitation of the soil can permanently alter the life-supporting properties of a forest or a once-fertile plain.

But, now, the size and persistence of environmental effects has grown with the power of modern science and the expansion of new technology. In the past, the environmental effects which accompanied technological progress were restricted to a small place and a relatively short time. The new hazards are neither local nor brief. Modern air pollution covers vast areas of the continents. Radioactive fallout from nuclear explosions is world wide. Synthetic chemicals have spread from the United States to Antarctica; some of them may remain in the soil for years. Radioactive pollutants now on the earth's surface will be found there for generations, and in the case of carbon-14, for thousands of years.

At the same time, the permissible margin for error has become very much reduced. In the development of steam engines a certain number of boiler explosions were tolerated as the art was improved. If a single comparable disaster were to occur in a nuclear power plant or in a reactor-driven ship near a large city, thousands of people might die, and a whole region rendered uninhabitable. Modern science and technology are simply too powerful to permit a trial-and-error approach.

This means that we cannot escape the responsibility of evaluating the competence of modern science and technology as a guide to human intervention in the environment. My own considered opinion is that modern science is a dangerously faulty foundation for technological interventions into nature. This becomes evident if we apply the so-called "engineering test" to it—that is, how well does it work in practice? Science represents our understanding of the natural world in which man must live.

Since man consciously acts on the environment through technology, the compatability of such action with human survival will, in turn, depend on the degree to which our technological practice accurately reflects the nature of the environment. We may ask, then, how successful is the understanding of nature which science now gives us as an effective guide to technological action in the natural world?

It is my contention that environmental pollution reflects the failure of modern science to achieve an adequate understanding of the natural world, which is, after all, the arena in which every technological event takes place.

The roster of the recent technological mistakes in the environment which have been perpetrated by the most scientifically advanced society in the history of man—the United States of 1969—is appalling:

• We used to be told that radiation from the fallout produced in nuclear tests was harmless. Only now, long after the damage has been done, we know differently. The bombs were exploded long before we had even a partial scientific understanding that they could increase the incidence of harmful mutations, thyroid cancer, leukemia, and congenital birth defects.

• We built the maze of highways that strangles almost every large city, and filled them with hordes of automobiles and trucks long before it was learned—from analysis of the chemistry of the air over Los Angeles—that sunlight induces a complex chain of chemical events in the vehicles' exhaust fumes, leading eventually to the noxious accumulation of smog.

• For more than 40 years massive amounts of lead have been disseminated into the environment from automotive fuel additives; only now has concern developed about the resultant accumulation of lead in human beings at levels that may be approaching the toxic.

• The insecticide story is well known: They were synthesized and massively disseminated before it was learned that they kill not only insects, but birds, and fish as well, and accumulate—with effects that are still largely unknown—in the human body.

• Billions of pounds of synthetic detergents were annually drained into U.S. surface water before it was learned—more than ten years too late—that such detergents are not degraded by bacterial action, and therefore accumulate in water supplies. Nor were we aware, until a few years ago, that the phosphates added to improve the cleansing properties of synthetic detergents would cause overgrowths of algae, which on their death pollute surface waters.

• In the last 25 years the amount of inorganic nitrogen fertilizer used on U.S. farms annually has increased about fourteen-fold. Only in the last few years has it become apparent that this vast elevation in the natural levels of soil nutrients has so stressed the biology of the soil as to introduce harmful amounts of nitrate into foods and surface waters.

• The rapid combustion of fossil fuels for power, and more recently, the invasion of the stratosphere by aircraft, are rapidly changing the earth's heat balance in still poorly understood ways. The outcome may be vast floods—or a new ice age.

• And, for the future, if we make the monumental blunder, the major military powers have prepared to conduct large-scale nuclear, chemical and biological warfare—which can only result, for belligerents and neutrals, in a vast biological catastrophe.

Each of these is a technological mistake, in which an unforeseen consequence has seriously marred the value of the undertaking. In order to illustrate the origin of such failures, I should like to discuss, briefly, the homely example of sewage disposal.

In natural lakes and rivers, animal organic wastes are degraded by the action of bacteria of decay which convert them into inorganic substances: carbon dioxide, nitrates, and phosphates. In turn these substances nourish plants, which provide food for the animals. In sunlight, plants also add to the oxygen content of the water and so support animals and the bacteria of decay. All this makes up a tightly woven cycle of mutually dependent events, which in nature maintains the clarity and purity of the water, and sustains its population of animals, plants, and microorganisms.

If all goes well, this biological cycle can assimilate added organic waste materials, and, maintaining its balance, keeps the water pure. But such a complex cyclical system, with its important feedback loops, cannot indefinitely remain balanced in the face of a steadily increasing organic load. Sufficiently stressed it becomes vulnerable at certain critical points. For example, the bacteria that act on organic wastes must have oxygen, which is consumed as the waste is destroyed. If the waste load becomes too high, the oxygen content of the water falls to zero, the bacteria die, the biological cycle breaks down, the purification process collapses, and the water becomes foul.

A sewage treatment plant domesticates the microbial activities that degrade wastes in natural streams and lakes. Sewage treatment involves a primary step in which indigestible solids are removed, and secondary treatment in a tank or pond rich in microbial decay organisms. During secondary treatment, the organic materials, artificially supplied with oxygen, are converted by microbial oxidation into inorganic substances. If the system works well, the resulting water is a clear, dilute solution of the inorganic products, of which nitrate and phosphate are most important. These inorganic products of sewage treatment, now free of oxygen demand, presumably can be released to rivers and lakes without causing any immediate drain on the oxygen in them.

But it has recently become apparent that this form of waste disposal technology is, to put it simply, a failure. For, in many places—for example, Lake Erie—the products of the treatment systems themselves, nitrate and phosphate, ultimately increase the organic load on the water, deplete the oxygen, and so negate the entire purpose of the system. Nitrate and phosphate are always present in natural waters—but in amounts far less than those generated by the huge waste load imposed on them by man. And at such abnormally high levels, nitrate and phosphate become a new hazard to the biological balance. These concentrated nutrients may induce a huge growth of algae—an algal "bloom." Such an abnormally dense population tends to die off with equal suddenness, again overloading the water with organic debris, and disrupting the natural cycle. To make matters worse, we are adding to the nitrate burden of surface waters by the massive use of nitrogeneous chemical fertilizers, and to their phosphate burden through the use of phosphate-rich detergents.

What all this means for the U.S. as a whole is evident from the report of the Spilhaus Committee report to the President's Office of Science and Technology. According to that report:

> The oxygen-demanding fraction of domestic and industrial waste is growing much more rapidly than the efficiency of waste treatment, so that, by 1980, it is estimated the oxygen demand of treated effluents will be great enough to consume the entire oxygen content of a volume of water equal to the dry-weather flow of all the United States' 22 river basins.

Ignorance of the biology of the environment is leading to the absolute deterioration of the quality of the nation's water systems. For example, despite a steady improvement in New York City's sewage treatment facilities, since 1948, in most of the coastal waters which receive the effluent of New York City treatment plants, the numbers of human intestinal bacteria has *increased* sharply. Water at many of the city beaches contains bacterial counts which are well above the allowable public health limits. Behind this astonishing fact may be a new and hitherto anticipated [sic] phenomenon—that intestinal bacteria, rather than dying off, as expected, when they are discharged from treatment plants into surrounding waters, actually multiply in them, because of the high concentration of bacterial nutrients, such as phosphate.

The warning is clear. We have begun to stress the self-purifying power of the Nation's surface waters to the point of biological collapse. A major cause of this impending catastrophe is our ignorance of the technological requirements of the environment and our persistent tendency to design technological instruments which do violence to these requirements.

The failure of technology is also evident in the air pollution problem. How else can we judge the matter of the recent proposal to construct a sunken expressway, capped with schools and houses, across lower Manhattan, without previous consideration of the resultant effects of carbon monoxide generated by the traffic on the school children and residents? Only *after* the plan was announced, did it become apparent that the resultant carbon monoxide levels would be sufficient to cause headaches, mental dullness, and even collapse. Or consider the potential impact of microscopic flakes of asbestos, spread from building materials and the lining of air conditioning ducts into the city air. Careful studies by Dr. I. J. Selikoff of the Mount Sinai School of Medicine of New York City residents show ". . . these particles are now very common among city dwellers at this time." And it has been established that the presence of asbestos particles in the human lung is the prelude, if sufficiently concentrated, to a particular form of lung cancer.

We tolerate the operational failure of the automobile and other technological hazards to the environment only because of a peculiar social and economic arrangement—that the high costs of such failures (for example, the lives lost to lung cancer or the medical cost of smog-induced emphysema) are not charged to any given enterprise, but are widely distributed in society. As a result, these costs become so intermingled with the costs due to other agents (for example, air pollution from power plants) as to become effectively hidden and unidentifiable. This suggests that the "success" of modern technology is largely determined by its ability to meet the economic requirements of the manufacturer. Measured against the economic interests of those who bear its costs in environmental deterioration—society as a whole—technology is by no means as successful.

In the same way, nuclear bombs epitomize the exquisitely refined control that has been achieved over nuclear reactions. There is no evidence that any U.S. nuclear bomb has failed to go off or to produce the expected blast. What has been unanticipated, and the source for loud complaint, is the effect of nuclear explosions on the biology of the environment. And again, the chemists and engineers who have given us the new synthetic detergents and insecticides are clearly competent to produce the desired materials; the trouble comes when we begin to use them, as they must be, in the environment.

No one can deny, of course, that, in certain respects, modern technology is brilliantly successful. Certainly, the modern mass-produced automobile is a technological triumph—up to a point. The dividing line between success and failure is the factory door. So long as the automobile is being constructed, it is a technological success. The numerous parts are designed, shaped, fitted together—and the whole assemblage works. However, once the automobile is allowed out of the factory into the environment, it is a shocking failure. It then reveals itself as an agent which has rendered urban air carcinogenic, burdened human bodies with nearly toxic levels of carbon monoxide and lead, embedded pathogenic particles of asbestos in human lungs, and has contributed significantly to the nitrate pollution of surface waters.

I believe that we must reverse the order of relationships which now connect economic need, technology, and the biology of the natural world. In the present scheme of things, narrow economic or political needs dictate the choice of a given technological capability—construction of nuclear weapons or an expressway, synthesis of insect-killing chemicals or the manufacture of asbestos building materials. This capability is translated into a specific engineering operation and when the operation is intruded upon the natural world, a myriad of biological problems arise—and it is left to the biologists, physicians, and others concerned with the survival of living things, to cope with these hazards as best they can.

I believe that if we are to assimilate modern science into a technology which is compatible with the environment that must support us, we shall need to reverse the present relationships among biology, engineering, technology, and economics. We need to begin with an evaluation of human needs and desires, determine the potential of a given environment to meet them, and *then* determine what engineering operations, technological processes, and economic resources are needed to accomplish these desires, in harmony with demands of the whole natural system.

We need to reassess our attitudes toward the natural world on which our technology intrudes. Among primitive people, man is seen as a dependent part of nature, as a frail reed in a harsh world, governed by immutable processes which must be obeyed if he is to survive. The knowledge of nature which can be achieved among primitive peoples is remarkable. The African bushman lives in one of the most stringent habitats on earth; food is scarce, water even more so, and extremes of weather come rapidly. The bushman survives because he has an intimate understanding of his environment.

We who call ourselves advanced claim to have escaped from this kind of dependence on the environment. Where the bushman must squeeze water from a searched-out tuber, we get ours by the turn of the tap. Instead of trackless wastes, we have the grid of city streets; instead of seeking the sun's heat when we need it, or shunning it when it is too strong, we warm ourselves and cool ourselves with man-made machines. All this tends to foster the idea that we have made our own environment and no longer depend on the one provided by nature. In the eager search for the benefits of modern science and technology, we have become enticed into a nearly fatal illusion: that we have at last escaped from our dependence on the balance of nature.

The truth is tragically different. We have become, not less dependent on the balance of nature, but more dependent on it. If we fail to understand this inescapable fact of modern life, we shall forfeit our survival.

We are still in a period of grace. In that time, let us hope, we can all learn what the architect has long known—that the proper use of technology is not to conquer the world, but, instead, to live in it.

INTRODUCTORY NOTE

Perry Miller's essay, "The Responsibility of Mind in a Civilization of Machines," provides a transition from the contemporary to the historical sections of this book. Miller wrote of attitudes that came out of America's early nineteenth century and persisted strongly a century later. He believed that Alexis de Tocqueville, the perceptive and articulate French visitor to America in 1832, erred in anticipating an attenuation of American utilitarianism stimulated by the exploitation of technology. By contrast, Miller found that as the gadgets multiplied Americans became "more and more bemused by the glory, the thrill of the technological transformation."

Miller insisted that the dream of the age in which Thoreau lived was not Walden Pond but J. A. Etzler's *Paradise Regained.* Americans in the nineteenth century—to use Miller's metaphor—were "grasping for the technological future, panting for it, crying for it. . . ." Tocqueville and many others after him could not and did not foresee or comprehend fully "the passion with which these people flung themselves into the technological torrent, how they shouted with glee in the midst of the cataract, and cried to each other as they went headlong down the chute that here was their destiny. . . ."

Miller, a scholar immersed in the popular literature of the nineteenth century, found little evidence that Americans anticipated that the telegraph, steamboat, Howe press, and similar devices considered mechanical slaves would become within a century their masters. In 1961, Miller saw technology in the saddle and Americans, although dimly perceiving the prick of the steel spurs, still unaware of the direction events had taken. In this essay, he expressed dismay at Americans' lack of awareness of their helplessness in a civilization of machines. The chaos of civilization and the inability of men to gain control of it alarmed him. Perhaps he would have sensed an increased awareness of the problem if he had lived into the seventies, for the writings of Mumford, Roszak, Commoner, and others were not only concerned, like Miller, but also drew public attention to such issues as environmental pollution and such problems as technological determinism, the inexorable onrush of technological change.

Perry G. E. Miller (1905–1963) was a professor of American literature at Harvard University and a brilliant student of American intellectual history who considered the mind as a maker of human history. His *The Life of the Mind in America* won the Pulitzer Prize and his *The New England Mind* was also an influential study. His other works include *Roger Williams, Jonathan Edwards,* and *Errand into the Wilderness.*

The Responsibility of Mind in a Civilization of Machines

PERRY MILLER

Ever since 1840, when Alexis de Tocqueville published the second volume of *Democracy in America,* which was quickly translated and avidly studied in the United States, anyone speaking formally about the role of intellect in our civilization is virtually forced to invoke Chapters IX and X of his First Book to explain why, to repeat his own chapter heading, "the Americans are more addicted to practical than to theoretical science."

We must be meticulous, therefore, to observe sanctified ritual by reciting, as though an incantation to the gods, at least a few of Tocqueville's venerable pronouncements. Certain of these are worth repeating when we recollect how immensely helpful they were to pioneer champions of the "theoretical" in the early struggles of American scientific ambition. In the long and bitter fight over the structure of the Smithsonian Institution, for instance—a battle which, by the ironic comedy of Smithson's leaving his bequest "to the United States of America," had to be fought, of all unlikely places, on the floors of Congress—Tocqueville proved a powerful aid to Joseph Henry, Alexander Bache and John Quincy Adams in their resistance to the Congressional attempts to furnish only a menial trade school. The victory, in 1846, of these eastern "theoreticians" over the grim-faced "utilitarians" from Ohio and Indiana is customarily saluted in chronicles of the American mind as a vindication of Tocqueville's bold surmise that the then observable state of American culture was not to be taken, considering the nation's youth and immaturity, as precluding in some distant future an appreciation of abstract or pure science among a democratic people.

Because I am convinced, after years of being bored with

Reprinted from *The American Scholar,* Volume 31, Number 1, Winter, 1961–1962.

the Tocqueville myth, that long citations from him are the handiest means by which social historians, when expounding American manners, evade intellectual responsibility, I am the more prepared to put before you these hoary articles as proceeding from faulty observation and naïvely *a priori* notions, and as being correct only in a limited range of matters, largely by accident. In this sense Tocqueville is a help to us, much more than if he had always been right. His confusions illuminate ours, as we here move in unorganized bands in the midst of a darkling plain, often striking at friends whom we do not recognize and at foes who in reality are our friends.

On the surface Tocqueville is highly persuasive. Let us supose, he wrote, that the Americans, coming to the wilderness with all the knowledge they actually did import, had then been left alone in the world, with no further importations from intellectual Europe. In that case, he asserted, "they would not have been slow to discover that progress cannot long be made in the application of the sciences without cultivating the theory of them." However absorbed they might have been with applications, Tocqueville strove to assure himself—if only because he was dismayed by their apparent fanaticism for the practical—that Americans would speedily admit the necessity of "theory," because, being a shrewd people, they would see in it a means of attaining the effectual end.

Or in other words—if I may venture to paraphrase Tocqueville—Americans were at that stage of their development exclusively prèoccupied with the useful arts and with the comforts of living. This was not too reprehensible when one reflected, as did Tocqueville, that "learned and literary Europe" was also striving to improve those means that ministered to the pleasures of mankind. But at the same time Europe was vigorously "exploring the common sources of truth." By contrast, in America all the incentives were toward utility. Whereupon Tocqueville came to the happily ambiguous conclusion that, while men in a democratic society are not inherently indifferent to the sciences and to the arts, they are obliged to cultivate them "after their own fashion and bring to the task their own peculiar qualifications and deficiencies."

Thus Tocqueville did not quite despair of the scientific mind in a democracy, yet neither was he overconfident about its future. During the short period of his visit—less than a year—

he spent a large part of his precious time with Whig lawyers in New York and Boston, especially with Mr. Justice Story, and with such pundits as Jared Sparks; by them he was supplied with several low estimates of the average American intelligence. These informants were so eager to explain to him the ignorance and barbarity of Andrew Jackson that they quite forgot to call his attention to the then mounting surge into technology.

Left to himself, this highly logical Frenchman was wondrously unqualified to comprehend what he saw. Hence, while he was generous enough to allow that a democracy, when it found the leisure to direct a particle of its energies to theory, might prove to have some qualifications for abstraction, he had also to stress its probable "deficiencies." Tocqueville was incapable of grasping how, in a democracy, traits that he termed deficiencies would not be admitted by the democratic scientist to be serious defects. Joseph Henry did contend at times during the contest over the Smithsonian, "He who loves truth for its own sake, feels that its highest claims are lowered and its moral influence marred by being continually summoned to the bar of immediate and palpable utility." Yet Henry himself was not averse to the useful; his experiments on electromagnetism and what he called "electrodynamic induction" were eminently addressed to the use of mankind. His campaign for the Smithsonian did not mean that he would not participate in the national exultation over the steamboat, the railroad, the telegraph (which he almost invented), the Hoe Press. Henry and his ilk did not feel they were demeaning themselves when they pleaded that their concern with a science not aimed at immediate application was really a strategic maneuver to gain, in the long run, still more efficient machines.

As the gadgets multiplied—steamboats, railroads, telegraph—the intellect of the Republic was more and more bemused by the glory, the thrill of the technological transformation. I suppose one may say that we are still hypnotized by the glitter of supermarkets, television, turnpikes, frozen foods, bras and girdles, and—especially when vacationing in Europe—by the sublimities of our plumbing. Yet we become sadly unhistorical when we assume, as several of our polemicists do, that the early Americans were a simple, ascetic and pious rural people who suddenly had their idyllic way of life shattered by a barrage of mechanical contrivances. The truth is, the national mentality was not caught unawares, not at all so rudely jolted as is generally supposed. There were of course, as there still are, rural backwaters, where the people clung to the simpler economy and there was a certain amount of folk resistance to the temptations of the machine. But on the whole, the story is that the mind of the nation flung itself into the mighty prospect, dreamed for decades of comforts that we now take for granted, and positively lusted for the chance to yield itself to the gratifications of technology. The machine has not conquered us in some imperial manner against our will. On the contrary, we have wantonly prostrated ourselves before the engine. Juggernaut seems by contrast an amateur contrivance; we have invented the superhighway, an impressively professional mechanism for mass slaughter.

In order to comprehend the mentality that still prevails in the America of the 1960's, we must rediscover that of a century or more ago, from which our own is directly descended. The code was even then in process of formulation. The economy was agricultural and a frontier still beckoned masses away from the cities, but then the code appeared. Immigrants quickly learned the improvised game, and proved themselves adept at profiting from it, whereas older families often failed (but not always) to grasp it in time. The pioneers conquered the forest and plowed up the plains in the hope that they too would benefit by it. They were not fleeing the machine; they were opening the areas in which it could operate.

This is to say that when Tocqueville paid the democracy what he thought was the compliment of foretelling a day when it might redeem itself from its base standards of value, the democracy itself was identifying its innermost being with the vibration of this triumphant utility. It would dismiss with a sneer European pretensions to a theoretical superiority as just another sign of the incurable corruption of Europe; it would consider any addiction to this superstition within the native community an un-American activity. Tocqueville, supposing the basic problem to be a conceptual one, simply could not comprehend the passion with which these people flung themselves into the technological torrent, how they shouted with glee in the midst of the cataract, and cried to each other as they went headlong down the chute that here was their destiny, here was the tide that would sweep them toward unending vistas of prosperity.

We today are still bobbing like corks in the flood, unable to get our heads high enough above the waves to tell whether there are any longer solid banks on either side or whether we have been carried irretrievably into a pitiless sea, there to be swamped and drowned. Those who are worried enough to seek glimpses of the receding horizons are those most likely to acquiesce in the charges of vulgarity and materialism levied against America by outside observers, from Mrs. Trollope and Charles Dickens to Simone de Beauvoir. In recent years we are apt lamely to murmur that the situation is not so bad as it used to be: witness the increased attendance at opera, ballet and concerts, the huge sale of classical records, the throngs attending exhibitions of paintings and the guided tours of school children through art museums. In academic institutions those who administer courses that in modern terminology are called the "humanities," while out of one side of their mouths complaining that budget-makers and foundations and the Federal Government provide millions for laboratories but only a few pennies for languages, will out of the other side admit to having lately been encouraged by an augmentation in the number of their graduate students.

Our disposition to treasure these miscellaneous manifestations of a taste for culture, of a publicly advertised thirst for beauty not yet obliterated by the weight of technology, has been accentuated by a tendency among students of the American past to emphasize the minorities who have protested here and there against the majority's infatuation with the machine. By blowing them up to a size out of all numerical proportion to the mass, we perpetuate the notion that the way for mind to survive in the midst of the glare of blast-furnaces and the shriek of jet-planes is to construct tiny oases of gentility around the library, the art museum, the concert hall. We repeat to ourselves Emerson's lines about things being in the saddle and riding mankind in order to encourage ourselves to buck against the pricks of metallic spurs. Possibly the cult of Edgar Allan Poe is not quite so legion as it was in my youth, but anthologies reprint his sonnet, "To Science," and students still sigh with a sympathy, even though fleeting, for its denunciation of science for having mangled the imagination, for its rhetorical questioning of why the poet should be required to love this vulture,

> or how deem thee wise,
> Who wouldst not leave him in his wandering
> To seek for treasure in the jewelled skies,
> Albeit he soared with an undaunted wing?

Although we read this verse by the light of an electric lamp, warmed in the dead of winter by an oil-burning furnace, having dined on a dish of frozen peas and been gratefully assured with the latest reading on our electrocardiograph that we are in no immediate danger, while we plan a trip for the morrow in a scientific fabrication propelled by gasoline, we not only palpitate with pity for poor Poe, but by sharing his anguish we enjoy the illusion of being entirely independent of all these daily conveniences. Yet in point of brute fact we cannot do anything about their hold over us, any more than we can liberate ourselves from the indignities of commuter trains, not to mention the obscenities of subways. Still, as long as we can respond to Poe we are able to assure ourselves that we have not become wholly rigidified in what Henry Miller defined as the "airconditioned nightmare" of the American way of life.

A more eloquent symptom of the dislocation between the sensitive mind and the confessedly insensitive environment in which the machines have corralled us is a steady enlargement of the popular regard for Henry Thoreau. To judge from the number of editions now in paperbacks and the volume of their sales, the vogue of Thoreau is a phenomenon that no modern Tocqueville should ignore. Yet, while these reprints are sold openly in drugstores, there is something clandestine about the transaction. By rights, while *Lady Chatterley's Lover* is patently a legitimate article of commerce, *Walden* ought to be bootlegged under the counter. It does seem clear that the appeal of Thoreau is not mainly to beatniks who have signed off from the reign of the machine, but to hundreds most abjectly enslaved by it. Thoreau appeals to those prisoners of megalopolis who from him gain at least a passing sight of blue sky. He keeps alive the flicker of an almost extinguished fire of the mind amidst piles of nonflammable steel and concrete—and chromium.

Most of us, I am sure, find something admirable, or at

least moving, in this surreptitious adherence to Thoreau. I rejoice when told that in the lower echelons of Wall Street there are young executives who, once they have contrived through the rush hour to reach their ranch-type homes in Scarsdale, mix a bit of Thoreau with their martinis. Also I am uplifted when informed that even in the higher echelons there are several collectors of Thoreauviana. I hope that in the practice of their secret vice, although they shall never be manumitted, they will reach beyond *Walden* to Thoreau's lesser known but gay review of an otherwise forgotten production by one J. A. Etzler, which was entitled *Paradise Regained,* and which Thoreau lampooned by heading his discussion, "Paradise (to be) Regained." Etzler was a German immigrant so wonderstruck (as were almost all of his sort) at the technological prospect offered by America that he extrapolated a future that is astonishingly similar to our present. He proposed to bring this paradise on earth within not a century but a decade: "If we have the requisite power for mechanical purposes, it is then but a matter of human contrivance to invent adapted tools or machines for application." Breathes there among us a man with soul so dead as not to feel a slight thrill of release upon hearing Henry's riposte: "Every machine, or particular application, seems a slight outrage against universal laws. How many fine inventions are there which do not clutter the ground?"

Yes, many of us will confess that our ground has become fearfully cluttered. But then we have to ask ourselves, with how many of the inventions would we willingly dispense? The telephone? Television? The airplane? Metrecal? In this interrogation consists the fundamental query for the mind of America. It is pressed upon us, on the one hand, by our Far Eastern friends, who find us sunk in materiality and incapable of understanding such an embodiment of spiritual power as Gandhi. On the other hand, the same question is forced upon us by those empires that have found their configuration in dogmatic materialism, and therefore ridicule us as hypocrites and sentimental evaders of reality. They accuse us of being mentally incapable of coping with the colossal mechanism we have devised. They make the issue terrifyingly clear: if the mind of this society is not competent to master the assembly line, the reactor, the computer, if we can endure these monsters only by snatching at impotent dreams of Walden Pond, then it is fitting that the machines take over; it is time that the mind abdicate. A recent authority—significantly remaining anonymous—has publicly asserted that with "perceptron" we shall shortly have a mechanical intelligence infinitely more advanced even than the computer, a true robot; then he adds, "But remember, all this was begun and devised by human brains, so humans—if they take care—will remain supreme." The nervous giggle that his cunning insertion, "if they take care," is designed to excite is all too evidently a momentary release of an anxiety steadily gnawing within us, the fear that possibly we shall not be able to take sufficient care. The essence of great tragedy, Eric Bentley says, is the realization by the self that it is totally unequipped to confront the universe. We might venture that even more tragic than any classical or Shakespearean drama is the crisis of illumination when man realizes, much too late for any last-minute panaceas, that he is unequal to the task of dealing with a universe of his own manufacture. Gloucester in *King Lear* blames the Gods who kill us for their sport, as wanton boys do flies. Shakespeare was even then at liberty to accuse the "Gods." But whom dare we blame for Gary, Indiana?

At this point it may be rewarding to consider more objectively than we have hitherto the mentality that in the period of Tocqueville's visit was welcoming the onrushing age of mechanization. In fact, "welcoming" is too pallid a verb: the age was grasping for the technological future, panting for it, crying for it. Such a reappraisal of the temper of the times becomes for us all the more urgent once we concede how seductively our fondness for Poe or for Thoreau distorts our image of the era. The critical historian must restrain his sympathies; in all candor he must report that the dissenters were at best minor voices and that they were sadly ineffectual. They provide us today with no usable programs of resistance. Whereupon we are compelled to pose the question we try to avoid. Is it, after all, the real issue that mind must stand in an attitude of intransigence against the machine? Is this opposition, celebrated by poets and encouraged by the heritage of romantic literature, this flattery of the private self—is this the one and only frame of reference in which the problem must be stated? Is it true that the mind in America is condemned always to be arrayed in unrelenting antagonism to the dynamo? If so, the outcome is inevitable: such a resistance movement is bound to be unavailing; if so, the mind is foredoomed to recurrent and increasingly disastrous defeat. From the commencement of the industrial transformation of the agrarian economy so dear to the heart of Thomas Jefferson, this doctrine of the mind, inherited from a thousand sources in Western history, from Christianity and also from John Locke, has collapsed. If this be all that the word "mind" can evoke in the modern situation, to what then have we to cling, as we find ourselves upon the wastes of matter, but the floating fragments of Poe's poetry, of Thoreau's individualism and odd bits and pieces of Henry Adams, Veblen and Randolph Bourne?

Or else must we cultivate what Thoreau scorned as resignation? Whereupon we admit that human existence is incurably shabby, that it is merely another evidence of the operation of inherent depravity, that the terrors of the machine can be withstood only by a delicious few anchored in religion. There is abundant evidence that to many this option appears the only way out. These anchorites will resent my describing their clause as one of escape. They insist that by their reinvigoration of spiritual resources the mind will be fortified against the thrusts of the material, will subdue the inhuman to humanity. However, the most energetic of modern Protestant thinking—I have in mind Reinhold Niebuhr's—hardly lets us off the hook. Catholicism does indeed have the majestic order of the historic Church; yet, so far as I can follow them, individual Catholics are as perturbed as the rest of us. Without entering further into a highly complex area, let me simply assert that an over-all examination of American history from about 1815 gives no support to any contention that the religious solution is workable for more than a few votaries of particular persuasions.

In the early nineteenth century the loud hosannas of the revivalists were as much swamped by the swelling chorus of praise for the machine as were the drawling sarcasms of Henry Thoreau. If we wish to locate where in the dominant theme of American culture the prospect of a healthy relation of mind to machine arose, and where it assumed the character that has subsequently ruled our society—disregarding the whimpers of humanists, religionists, agrarians and poets—we must pass over names of literary or ecclesiastical repute. We are obligated, for instance, to resurrect an address of Charles J. Ingersoll to the American Institute of Philadelphia in November, 1835:

> Even in Europe . . . this is the era of utility. With us it is the age of navigation, plantation, habitation, and transportation, of steamboats, canals, and railroads, magnificent prisons, costly poor-houses, school-houses, custom-houses, court-houses, warehouses, factories, forges, laboratories, and all the contrivances of ingenuity.

If this does not adequately convey wherein the majority of Americans conceived their future to lie, let me further illustrate the lesson by repeating what Henry Meigs proclaimed upon the opening of the Annual Fair at Albany in 1850:

> All over our great domain we hear the ceaseless hum of human and machine labor. The latter has become in our time the object of wonder. We are almost as much astonished at modern inventions, as our Indians were at the ships and artillery of Columbus. By the constant habit of observation, and with entire self-reliance, and with a liberty which has no other boundaries than those which morality and religion impose, the freemen of our country will carry to the uttermost perfection all the arts that can be useful or agreeable to man.

Now, as I trust is clear, I am not arguing that in the record of creative intellect Henry Meigs is to be placed on a level with Henry Thoreau. Nevertheless, the fact remains: Meigs spoke for the community, and Thoreau did not. Thoreau spoke for himself, but not for Pittsburgh.

Further, men like Meigs were not uncultured. They were certain that when orating in this vein they were not becoming Philistine barbarians. They marvelled at the machine, but they took full responsibility. Never did any weird notion that the machine might someday dominate the men cross their minds. There was indeed some uneasiness throughout the country—not confined to "radical" circles—that the workers (or the "operatives" as was then the word) might be victimized. But this resulted in a determination that American industry had no need to follow the course of the English, that it resolutely should not. Thoreau again spoke the extreme concern:

> The condition of the operatives is becoming every day more like that of the English; and it cannot be wondered at, since,

> as I have heard or observed, the principal object is, not that mankind may be well and honestly clad, but, unquestionably, that the corporations may be enriched.

Again, Thoreau alerts our attention, where Meigs merely bores us. But the Meigses of his time (I suppose plenty of them survive in today's Chambers of Commerce) would have seen in Thoreau's passage, had they ever noticed it, only further proof of Thoreau's irresponsibility. Emerson also had moments of asking whether the machines might not damage the workers; but he, unlike Thoreau, was so in tune with his times as to be able to banish such apprehensions and to say in resounding amplitude, "Machinery and Transcendentalism agree well. Stage-Coach and Railroad are bursting the old legislation like green withes." His essay "Napoleon" in *Representative Men,* although at the end it brings to bear the same pompous moral judgment on the man of action that finds Shakespeare guilty of frivolity, is in substance his love letter to the entrepreneurs, to the practical men who brushed aside the "old legislation" and were building railroads.

Likewise Theodore Parker, the professional reformer pitching into every cause that offered itself, was worried about the condition of the laboring classes; however, he even more than Thoreau was the Yankee handy with tools, and to him the machines in themselves were as much objects of sheer beauty as to Francis Cabot Lowell. Parker hailed the factory as a creation of intellect. The observer of a carpet manufactory, he said, departs "wondering, thinking what a head it must be which planned the mill, a tool by which the Merrimack transfigures wool and dye stuff into handsome carpets, serviceable for chamber, parlor, staircase, or meeting-house." Occasionally the Old Testament prophet strain in Parker would predict that just so soon as the machine threatened to afford workers the leisure for intellectual relaxation, the bosses would invent markets for useless luxuries in order to keep their employees tied to a fourteen-hour-a-day regime. Yet these utterances came easily when he was tongue-lashing the Boston industrialists, several of whom came to his meetings in the Melodeon and helped pay his salary in order to be lashed. When, by contrast, Parker wanted to array the strength and virtue of his region against what for him was the supremely monstrous evil of the century, Negro slavery, he had no hesitation in telling Southerners that they, hamstrung by this iniquity, could never develop the exquisite machines that made New England almost such a paradise as Etzler had foretold:

> While South Carolina has taken men from Africa, and made them slaves, New England has taken possession of the Merrimack, the Connecticut, the Androscoggin, the Kennebeck, the Penobscot, and a hundred smaller streams. She has caught the lakes of New Hampshire, and holds them in thrall.

Thus inspired, Parker would promise that "the machinery" of the free states would outproduce the three million bondsmen of the South. Although he, dying of exhaustion in 1860, did not live to see the Civil War prove his thesis, hundreds of his admirers, those who did not perish in the demonstration, did so perceive it and pointedly included his sagacity in their encomiums. In those respects the reformer Parker was not at all out of step with what some historians term the "conservative" temper of his age. He would fully have applauded, as would have all except a few grumblers like Thoreau, a paragraph that another Meigs, this one J. Aitken Meigs, spoke in Philadelphia on July 13, 1854:

> Such is the general picture; examine it more carefully for a moment. See how true children of the sun, heat, light and electricity have been delivered into the hands of man, as bondservants, obedient to his call. . . . With consummate skill the marriage of water and heat was effected. The child of that marriage has grown to be a herculean aid to onward-moving humanity. Certainly steam is a benefactor to the race. The printing press and the electric telegraph have become the handmaids of thought.

"Bondservants!" "Handmaids!" There we have it, the veritable American religion! Such assertions that machines had become creatures of the mind were customarily followed, as with Meigs, by the substantiating proposition that in turn they would obligingly prove to be assistants to the intellect. They would further the empire of "intellect," they would add trophies to its dominion, they would be dutiful and obedient. As early as 1817, for instance, when the industrial revolution in the northeastern states had barely commenced, a society was formed for the encouragement of domestic manufactures, not, according to its own profession, because of the assurance of profit (as Henry Thoreau would insinuate), but in the cause of intellectual fulfillment. "The exhaustless stores of mind and matter shall be this nation's treasury."

When therefore, we must ask ourselves, do glowing periods in this optimistic strain begin to excite in our breast only a languid pity for those simple decades, or else a bitter resentment at having been so cheaply betrayed? Why do I discover when addressing a classroom loaded with the heirs of industry and with future vice presidents that the mordant aphorisms of Thoreau are greeted with an appreciative recognition, while the prognostications of industrial bliss are heard with an obvious subsiding of enthusiasm? Surely it is not that these youths are actively in revolt against the machine, like the rick-burners of Wellington's England. On the contrary, we all live by means of several hundred more apparatuses, of increasingly vaster scale, than either of the Meigses could have imagined. And nobody doubts that these instruments require intelligence to design, intelligence to construct and even more intelligence to improve. The Massachusetts Institute of Technology is a citadel of the mind in America, if any bastion is. Whence, then, comes this insecurity, which my title assumes is everywhere recognizable? Where, when and above all how did the mind of this fabulously successful enterprise become darkened with a dread that its jinni—summoned not out of a magic bottle but from a supposedly rational conjunction of the test tube, the wind tunnel, the steam shovel and sweet Professor Einstein's equation—would become its mortal enemy?

The most obvious answer is that millions of Americans, more than enough to win an election, have only vague notions, barely restive worries, as to the existence of any such enmity. The less obvious but more terrible fact is that they have no slightest sense of responsibility for any bifurcation of which they cannot conceive. They dwell in a mental fog of perpetual neutralism. They remain there despite reports of children burned to death in what newspapers euphemistically call "tenements," despite the disintegration of urban transportation, despite the increase of murders and disfigurements on our highways, despite the criminal pollution of the very air we breathe. Yet after countless tirades on dangers such as these have left the public staring in blank incomprehension, we further have to confront, as have many frustrated scientists, the massive indifference of the populace, most of them supplied with a standard set of human emotions, to the ultimate meaning of a war conducted through the mysteries of nuclear physics.

Can it be that a father is incapable of grief for the daughter miserably crushed in an "accident" on Route 999? Do children in a burning tenement not suffer all the agonies of Joan of Arc? And we all have been told, by John Hershey and by the Bombing Survey, what an atomic blast does to our love-life, our libraries, our galleries, our universities, to our fragile oases as well as to military installations. Except that in this department we are reliably informed that Hiroshima was child's play; the next bomb will obliterate consciousness itself. We have no choice but to consent, and so go dumbly down the appointed runways as do cattle in the stockyards.

How did the disseverance come about? This question is widely debated among historians, sociologists, psychologists and the more conscientious of physicists. In American historiography the fashion has become to regard *The Education of Henry Adams* as the tocsin of alarm, even though none heard any fire bell when the book was first printed in a private edition in 1907. When it was released to the public, in 1918, it surprisingly became a best-seller, possibly indicating that it by then found a partially disturbed audience. Yet the relation of the work to the current frame of the public mind is a baffling business. My late colleague, Thomas Reed Powell of the Harvard Law School, used to growl that the title of Adams' autobiography ought, in loyalty to its content, to be "The Unfortunate Relation of the Universe to Me." If this be so, and many equally exasperated with Adams' conceit and attitudinizing think it is so, the *Education* is the reverse of Mr. Bentley's definition of tragedy. Therefore it has upon the mind of the present, especially upon the minds of students for whom Adams is prescribed reading in college surveys, a dulling instead of an arousing effect.

Let us grant all this. It remains a fact that Adams endeavored to concentrate on the dynamo as the master engine—the archengine, shall we say?—of the twentieth century. But in the operation of a dynamo there remains a vestige of human responsibility: at least the engineer can, when it threatens to get out of hand, pull a switch. Yet Adams was sensitive enough, or neurotic enough, to read through the dynamo to the void beyond it: "In plain words, Chaos was the law of nature, Order was the dream of man."

I know some will object, and I heartily go along with their insistence, that a contemporary physicist so thoroughly trained in his métier as to be able still further to "improve" our bombs must be convinced that he manipulates not a chaos but some sort of order. This realm is so minuscule, so imperceptible by even the most powerful of microscopes, that it can be calculated only by a sign language that is a scandal to traditional norms of human reason; even so it has, according to these high priests, enough coherence of its own to enable their minds, if not ours, to produce tangible results in more and more lethal explosions.

This may be the insoluble conundrum of the epistemology with which we have to live—or perish. It would work extremely well if we were medieval serfs allowed an occasional glimpse of a Mass, the mysteries of which we were assured were beyond our comprehension. The odd circumstance is that in our case the hierophants have striven to divest themselves of the mystery, to account for themselves in the language of laymen, and have subsided into sorrow when obliged to confess that they cannot interpret physics to any but fellow technicians. Henry Adams was brought to the abyss by a sentence of Karl Pearson's: "In the chaos behind sensations, in the 'beyond' of sense-impressions, we cannot infer necessity, order or routine, for these are concepts formed by the mind of man on this side of sense-impressions." If at the beginning of our century—now seemingly as remote as the age of Voltaire—a responsible scientist warned the common-sense intelligence that it could not follow these lines of investigation past the barriers of sense impressions, and so would have to crouch outside the gates, hugging its cold and empty jars of order and law, how much more crushing is the exclusion when a J. Robert Oppenheimer says flatly that the working principles of modern physics cannot be translated into the discourse of logic, or when a C. P. Snow concludes that the scientists and the humanists can no longer speak a common language! If this has been the true, and the deceptive, course of history in our time, only one conclusion is left: the mind, as our Western culture has conceived it since the antiquity of Thales, is cut adrift. It is not even shut up, as Emerson feared Kantian Idealism would confine it, in the splendid labyrinth of its subjective perceptions. It can behold splendors and colors to its heart's content; but that is an irrelevant game, less related to the actual universe than whist or baseball. There is no point in rigorously subjecting the self to the ascetic discipline of Walden Pond; nothing is to be gained that way any more than through an afternoon at the stadium. The mind, as men in general know it, and through which they have tried to manage their universe, is no longer responsible for anything, least of all for the destruction of its own species.

This attitude, somehow communicated to the least reflective of our population, shows itself repeatedly in our dealings with our machines. Everyone who starts a drive in his automobile assumes that the collision will not happen to him. Otherwise he would never take his car out of the garage. The commuter notes in the headlines of his morning papers the number of the dead in a tenement fire and turns hastily to the sports page, meekly grateful that he has so much as a train to ride in. Yet within these areas the assuring consolation persists that even the average mind of the citizenry *could*, if worse came to worst, learn the mechanism of the internal combustion engine and so mitigate the holocausts on the roads during holiday weekends. The collective intelligence *might* frame and see enforced decent zoning laws. The fact that these attainments are as yet neglected by the great American public seems no reason for accusing the machines. I am told that a Broadway producer, having read a script about the destruction of Sodom and Gomorrah submitted to him by an aspiring young playwright, mused that he did not see how there could have been so much sin in communities that had no telephones. Here we are in a field where the invention itself does not have to assume full responsibility for human fallibility.

But if in an age of machines and of helpful gadgets our propensity be nourished to live with less and less understanding of all that we ought to comprehend, what happens when our debilitated faculty is told that it has to live under the shadow of nuclear weapons that by their very nature defy the few lingering canons of rationality?

Virtually all reports on the general behavior of Americans add up, so far, to a pattern of further and further regression into the womb of irresponsibility. There is everywhere documented a refusal to accept what I would hopefully term adult status. I shall construct a dialectic too simplified to suit any social scientist, but roughly it appears to run something like this. First, because there is nothing this or that particular individual can do to prevent the bombs from falling, then, if they do fall, the fault is none of his. Although they be launched by man-made missiles or dropped from man-made jets, and although man be exterminated, he remains morally immune, an innocent victim of the machine. Second, if, as several analysts assure us, the threat of mutual obliteration will itself keep the bombs from falling—as it prevented the use of poison gas in the last war—then our citizen can also claim that the fault is none of his. These may be the sheer alternatives with which we are confronted; there would seem to be no third recourse.

Well, that predicament is dire enough. But I see how no one, in the light of the historical development that I have sketchily indicated and in which, as I have endeavored to suggest, we are all involved, can with self-respect complain about the situation in which we find ourselves, since we have all done our utmost, knowingly or inadvertently, to produce it.

Obviously we cannot turn the clock back to the idyllic industrialism of Theodore Parker. What, then, can we say? We may say that without recourse to romantic isolationism we are able to resist, and will resist, the paralyzing effects upon the intellect of the looming nihilism of what was formerly the scientific promise of mechanical bliss. Scientists and humanists are today joined in their appreciation of the urgencies, and of the difficulties, of the consultation. Upon all of us, whoever we be, rests the responsibility of securing a hearing from an audience either dazzled into dumb amazement by the prestige of technology or else lulled into apathy by the apparently soothing but actually insidious triumphs of functional science.

Supposing that my analysis has any validity, then at the end of it we come, by cumbersome indirection, to the problem of the officially designated "humanities" in a civilization conducted amidst machines. I have tried not to put this cause in the center of the picture, for, as I have suggested, it does not belong there. Most of the keening over administrative slighting of the humanities has as little relevance to the main stream of historical movement as did Thoreau's magnificent tirades from the shore of Walden Pond. Except, of course, that not much of the modern complaint reaches Thoreau's eloquence or wit. Nothing is to be gained by acknowledging in advance a purely ornamental function, or by proclaiming an inutility for which compensation is to be found in "poetry," in order to demand a shelter for the arts within a capitalist order.

The Cassandras of our time have exaggerated the danger when they cry, "Woe! woe! the scientists and the humanists can no longer converse!" Sir Charles Snow is a salutary sort of Cassandra, because he is not hysterical, and furthermore is living proof that the dialogue can be conducted. So a conversation is possible; in the basic sense, the humanist and the scientist are both on the same side of the barricade. It is not required every time a historian and a chemist meet for lunch that each impart to the other the sum and substance of his erudition. It is only required that they respect in each other an achievement of "mind." Baseball is endlessly fascinating, but it need not be the sole noncontroversial diversion when intellect meets intellect.

If such reflections have relevance for conclaves of scholars, they have an even greater importance for our relations to the public, albeit the amorphous, mind. In this direction the scientists, or at least the students of science, have lately been doing the better job. Admit that they have an immense prestige to support them, and that in a paradoxical sense the incommunicableness of their mystery is an incentive for attempting to communicate; they have addressed the public in various degrees of agonized clarity and, to that extent, they have builded dikes against the flux of irresponsibility. By contrast the humanities have, I must confess, not functioned so effectively, despite the media offered by FM radios and by paperbacks. Under the large rubric of humanities, in this indictment, I include not only writers on history and literature, but social scientists, political scientists and even, although grudgingly, economists. Despite the protest of many of these that they do communicate with the public, it is in such areas that the analysis of the mind in America has itself been undergoing a fragmentation into theses and monographs. These productions are intellectually respectable, but too often they add up to a tiresome jumble. They become a disease, and the endeavor to check the disease oddly enough turns into an agent of contagion.

When humanists complain that vulgarizers pre-empt the market, they must also admit that they have to blame their cumbersome styles, the dismal catalogues that they substitute for perceptions, and their insensitivity to the living concerns of the populace. They prove themselves as unheeding as are those of whose unheedingness they wail. They too affect an innocence, behind which they conceal their inability to assume responsibility in the world of mechanical stress. Yes, there are subjects everywhere within the "humanities" that cannot be popularized, in which no such effort should be attempted. But where is the line of separation? Only in the mind that can address with respect another mind.

It may be that every exertion toward a community of learning is bound for a long time, for the remainder of our existence, to come to no decisive conclusion. But for us, is success the only goal? Upon those who have the concern, and to match it the energy, rests the self-assumed burden of responsibility. Having accepted the hospitality of the obligation, we can not shake off the consequence. Yet, is it nothing but a burden? Is it only a dead weight we have sullenly to carry about the streets? Is it merely an ordeal to which we must submit? Not in the least. Like the precious, beautiful, insupportable and wholly irrational blessing of individuality, with all the myriad quandaries of responsibility therein involved, the responsibility for the human mind to preserve its own integrity amid the terrifying operations of the machine is both an exasperation and an ecstasy.

PART II

The Thrill of the Technological Transformation

INTRODUCTORY NOTE

"Paradise (To Be) Regained," signed "T," is a review by Henry David Thoreau of a book by J. A. Etzler, a "German immigrant" thought by Perry Miller to be "so wonder-struck (as were almost all of his sort) at the technological prospect offered by America that he extrapolated a future that is astonishingly similar to our present." Etzler's enthusiasms were representative of those of the Americans who, Miller observed, "positively lusted for the chance to yield . . . to the gratification of technology." Etzler took new modes of power, transportation, and machinery (described in some detail in Thoreau's review) as the means of recovering the idyllic state not known since the expulsion from Eden: "I promise to show the means of creating a paradise within ten years, where everything desirable for human life may be had by every man in superabundance, without labor, and without pay." He also cherished technology as the means for man to transform the face of the earth: "The whole face of nature shall be changed into the most beautiful forms." Like Goethe's *Faust,* written not long before his own, Etzler's book conceived of Western man's creative drive as transcendent. Like Faust, the Americans would drain the most dismal swamp, fill and level it, and create a habitable land laced with canals and graced by gardens. (Faust's land reclamation project nearly caused him to lose his soul to Mephistopheles, but Etzler foresaw no such danger for Americans.)

Thoreau, having conveyed the essence of Etzler's program, then asked if it could be fulfilled. His conviction that the spiritual rather than the physical landscape demanded cultivation and that this could be done only by the power of love brought him to answer "no." He had seen too many men fail to find the sustaining faith and the spirit of cooperation that the fulfillment of Etzler's utopian dream assumed; Thoreau also had found that a proliferation of machines, a manipulation of the external environment, did not bring order, serenity, and dignity. "There is a speedier way," Thoreau wrote, "than the Mechanical System can show to fill up marshes, to drown the roar of waves, to tame hyenas, secure agreeable environs, diversify the land, and refresh it with 'rivulets of secret water,' and that is by the power of rectitude and true behavior." Etzler would redeem the soil; Thoreau, the soul.

Thoreau found Etzler's vision flawed not only by his inattention to the world within, but also—the careful reader will note—by his disinterest in relating to nature on her own terms. Nature was to be overwhelmed by technology. Furthermore, Thoreau held that Etzler's inventions using the energy of the sun or the force of the wind would not contribute as much to man's well-being as the sunshine falling "on the path of the poet or the winds fanning his cheek." Again, Thoreau—like some mid-twentieth-century critics—balanced soul and nature against technology on the scale of human fulfillment and found the latter wanting.

Henry David Thoreau (1817–1862), essayist, poet, and transcendentalist, finished *Walden, or Life in the Woods* in 1853. For several years before the review essay on Etzler's book was printed, Thoreau had been living at the Concord, Massachusetts, home of Ralph Waldo Emerson and had become acquainted with a group known as the Transcendental Club. The philosophy of that group, its attachment to the deep significance to be discovered in seemingly simple natural things, like clouds or a flower, can be found in his Etzler essay. His rejection of the superficialities, commercialism, and materialism of the man-made world, so memorably conveyed in *Walden,* can also be found here.

*Paradise (To Be) Regained**

"T" (HENRY DAVID THOREAU)

We learn that Mr. Etzler is a native of Germany, and originally published his book in Pennsylvania, ten or twelve years ago; and now a second English edition, from the original American one, is demanded by his readers across the water, owing, we suppose, to the recent spread of Fourier's doctrines. It is one of the signs of the times. We confess that we have risen from reading this book with enlarged ideas, and grander conceptions of our duties in this world. It did expand us a little. It is worth attending to, if only that it entertains large questions. Consider what Mr. Etzler proposes:

> "Fellow Men! I promise to show the means of creating a paradise within ten years, where everything desirable for human life may be had by every man in superabundance, without labor, and without pay; where the whole face of nature shall be changed into the most beautiful forms, and man may live in the most magnificent palaces, in all imaginable refinements of luxury, and in the most delightful gardens; where he may accomplish, without labor, in one year, more than hitherto could be done in thousands of years; may level mountains, sink valleys, create lakes, drain lakes and swamps, and intersect the land everywhere with beautiful canals, and roads for transporting heavy loads of many thousand tons, and for travelling one thousand miles in twenty-four hours; may cover the ocean with floating islands movable in any desired direction with immense power and celerity, in perfect security, and with all comforts and luxuries, bearing gardens and palaces, with thousands of families, and provided with rivulets of sweet water; may explore the interior of the globe, and travel from

United States Magazine and Democratic Review, 13 (November 1843), 451–463.

*The Paradise within the Reach of all Men, without Labor, by Powers of Nature and Machinery. An Address to all intelligent Men. In two parts. By J. A. Etzler. Part First. Second English Edition. pp. 55. London, 1842.

> pole to pole in a fortnight; provide himself with means, unheard of yet, for increasing his knowledge of the world, and so his intelligence; lead a life of continual happiness, of enjoyments yet unknown; free himself from almost all the evils that afflict mankind, except death, and even put death far beyond the common period of human life, and finally render it less afflicting. Mankind may thus live in and enjoy a new world, far superior to the present, and raise themselves far higher in the scale of being."

It would seem from this and various indications beside, that there is a transcendentalism in mechanics as well as in ethics. While the whole field of the one reformer lies beyond the boundaries of space, the other is pushing his schemes for the elevation of the race to its utmost limits. While one scours the heavens, the other sweeps the earth. One says he will reform himself, and then nature and circumstances will be right. Let us not obstruct ourselves, for that is the greatest friction. It is of little importance though a cloud obstruct the view of the astronomer compared with his own blindness. The other will reform nature and circumstances, and then man will be right. Talk no more vaguely, says he, of reforming the world—I will reform the globe itself. What matters it whether I remove this humor out of my flesh, or the pestilent humor from the fleshy part of the globe? Nay, is not the latter the more generous course? At present the globe goes with a shattered constitution in its orbit. Has it not asthma, ague, and fever, and dropsy, and flatulence, and pleurisy, and is it not afflicted with vermin? Has it not its healthful laws counteracted, and its vital energy which will yet redeem it? No doubt the simple powers of nature properly directed by man would make it healthy and paradise; as the laws of man's own constitution but wait to be obeyed, to restore him to health and happiness. Our panaceas cure but few ails, our general hospitals are private and exclusive. We must set up another Hygeian than is now worshipped. Do not the quacks even direct small doses for children, larger for adults, and larger still for oxen and horses? Let us remember that we are to prescribe for the globe itself.

This fair homestead has fallen to us, and how little have we done to improve it, how little have we cleared and hedged and ditched We are too inclined to go hence to a "better land," without lifting a finger, as our farmers are moving to the Ohio soil; but would it not be more heroic and faithful to till and redeem this New-England soil of the world? The still youthful energies of the globe have only to be directed in their proper channel. Every gazette brings accounts of the untutored freaks of the wind—shipwrecks and hurricanes which the mariner and planter accept as special or general providences; but they touch our consciences, they remind us of our sins. Another deluge would disgrace mankind. We confess we never had much respect for that antediluvian race. A thorough-bred business man cannot enter heartily upon the business of life without first looking into his accounts. How many things are now at loose ends. Who knows which way the wind will blow to-morrow? Let us not succumb to nature. We will marshal the clouds and restrain the tempests; we will bottle up pestilent exhalations, we will probe for earthquakes, grub them up; and give vent to the dangerous gases; we will disembowel the volcano, and extract its poison, take its seed out. We will wash water, and warm fire, and cool ice, and underprop the earth. We will teach birds to fly, and fishes to swim, and ruminants to chew the cud. It is time we had looked into these things.

And it becomes the moralist, too, to inquire what man might do to improve and beautify the system; what to make the stars shine more brightly, the sun more cheery and joyous, the moon more placid and content. Could he not heighten the tints of flowers and the melody of birds? Does he perform his duty to the inferior races? Should he not be a god to them? What is the part of magnanimity to the whale and the beaver? Should we not fear to exchange places with them for a day, lest by their behavior they should shame us? Might we not treat with magnanimity the shark and the tiger, not descend to meet them on their own level, with spears of sharks' teeth and bucklers of tiger's skin? We slander the hyaena; man is the fiercest and cruelest animal. Ah! he is of little faith; even the erring comets and meteors would thank him, and return his kindness in their kind.

How meanly and grossly do we deal with nature! Could we not have a less gross labor? What else do these fine inventions suggest,—magnetism, the daguerreotype, electricity? Can we not do more than cut and trim the forest,—can we not assist in its interior economy, in the circulation of the sap? Now we work superficially and violently. We do not suspect how much might be done to improve our relation with animated nature; what kindness and refined courtesy there might be.

There are certain pursuits which, if not wholly poetic and true, do at least suggest a nobler and finer relation to nature than we know. The keeping of bees, for instance, is a very slight interference. It is like directing the sunbeams. All nations, from the remotest antiquity, have thus fingered nature. There are Hymettus and Hybla, and how many bee-renowned spots beside? There is nothing gross in the idea of these little herds,—their hum like the faintest low of kine in the meads. A pleasant reviewer has lately reminded us that in some places they are led out to pasture where the flowers are most abundant. "Columella tells us," says he, "that the inhabitants of Arabia sent their hives into Attica to benefit by the later-blowing flowers." Annually are the hives, in immense pyramids, carried up the Nile in boats, and suffered to float slowly down the stream by night, resting by day, as the flowers put forth along the banks; and they determine the richness of any locality, and so the profitableness of delay, by the sinking of the boat in the water. We are told, by the same reviewer, of a man in Germany, whose bees yielded more honey than those of his neighbors, with no apparent advantage; but at length he informed them that he had turned his hives one degree more to the east, and so his bees, having two hours the start in the morning, got the first sip of honey. Here, there is treachery and selfishness behind all this; but these things suggest to the poetic mind what might be done.

Many examples there are of a grosser interference, yet not without their apology. We saw last summer, on the side of a mountain, a dog employed to churn for a farmer's family, travelling upon a horizontal wheel, and though he had sore eyes, an alarming cough, and withal a demure aspect, yet their bread did get buttered for all that. Undoubtedly, in the most brilliant successes, the first rank is always sacrificed. Much useless travelling of horses, *in extenso,* has of late years been improved for man's behoof, only two forces being taken advantage of,—the gravity of the horse, which is the centripetal, and his centrifugal inclination to go a-head. Only these two elements in the calculation. And is not the creature's whole economy better economized thus? Are not all finite beings better pleased with motions relative than absolute? And what is the great globe itself but such a wheel,—a larger tread-mill,—so that our horse's freest steps over prairies are oftentimes balked and rendered of no avail by the earth's motion on its axis? But here he is the central agent and motive power; and, for variety of scenery, being provided with a window in front, do not the ever-varying activity and fluctuating energy of the creature himself work the effect of the most varied scenery on a country road? It must be confessed that horses at present work too exclusively for men, rarely men for horses; and the brute degenerates in man's society.

It will be seen that we contemplate a time when man's will shall be law to the physical world, and he shall no longer be deterred by such abstractions as time and space, height and depth, weight and hardness, but shall indeed be the lord of creation. "Well," says the faithless reader, " 'life is short, but art is long;' where is the power that will effect all these changes?" This it is the very object of Mr. Etzler's volume to show. At present, he would merely remind us that there are innumerable and immeasurable powers already existing in nature, unimproved on a large scale, or for generous and universal ends, amply sufficient for these purposes. He would only indicate their existence, as a surveyor makes known the existence of a waterpower on any stream; but for their application he refers us to a sequel to this book, called the "Mechanical System." A few of the most obvious and familiar of these powers are, the Wind, the Tide, the Waves, the Sunshine. Let us consider their value.

First, there is the power of the Wind, constantly exerted over the globe. It appears from observation of a sailing-vessel, and from scientific tables, that the average power of the wind is equal to that of one horse for every one hundred square feet. "We know," says our author—

> "that ships of the first class carry sails two hundred feet high; we may, therefore, equally, on land, oppose to the wind surfaces of the same height. Imagine a line of such surfaces one mile, or about 5,000 feet, long; they would then contain 1,000,000 square feet. Let these surfaces intersect the direction of the wind at right angles, by some contrivance, and receive, consequently, its full power at all times. Its average power being

> equal to one horse for every 100 square feet, the total power would be equal to 1,000,000 divided by 100, or 10,000 horses' power. Allowing the power of one horse to equal that of ten men, the power of 10,000 horses is equal to 100,000 men. But as men cannot work uninterruptedly, but want about half the time for sleep and repose, the same power would be equal to 200,000 men. . . . We are not limited to the height of 200 feet; we might extend, if required, the application of this power to the height of the clouds, by means of kites."

But we will have one such fence for every square mile of the globe's surface, for, as the wind usually strikes the earth at an angle of more than two degrees, which is evident from observing its effect on the high sea, it admits of even a closer approach. As the surface of the globe contains about 200,000,000 square miles, the whole power of the wind on these surfaces would equal 40,000,000,000,000 men's power, and "would perform 80,000 times as much work as all the men on earth could effect with their nerves."

If it should be objected that this computation includes the surface of the ocean and uninhabitable regions of the earth, where this power could not be applied for our purposes, Mr. Etzler is quick with his reply—"But, you will recollect," says he, "that I have promised to show the means for rendering the ocean as inhabitable as the most fruitful dry land; and I do not exclude even the polar regions."

The reader will observe that our author uses the fence only as a convenient formula for expressing the power of the wind, and does not consider it a necessary method of its application. We do not attach much value to this statement of the comparative power of the wind and horse, for no common ground is mentioned on which they can be compared. Undoubtedly, each is incomparably excellent in its way, and every general comparison made for such practical purposes as are contemplated, which gives a preference to the one, must be made with some unfairness to the other. The scientific tables are, for the most part, true only in a tabular sense. We suspect that a loaded wagon, with a light sail, ten feet square, would not have been blown so far by the end of the year, under equal circumstances, as a common racer or dray horse would have drawn it. And how many crazy structures on our globe's surface, of the same dimensions, would wait for dry-rot if the traces of one horse were hitched to them, even to their windward side? Plainly, this is not the principle of comparison. But even the steady and constant force of the horse may be rated as equal to his weight at least. Yet we should prefer to let the zephyrs and gales bear, with all their weight, upon our fences, than that Dobbin, with feet braced, should lean ominously against them for a season.

Nevertheless, here is an almost incalculable power at our disposal, yet how trifling the use we make of it. It only serves to turn a few mills, blow a few vessels across the ocean, and a few trivial ends besides. What a poor compliment do we pay to our indefatigable and energetic servant!

> "If you ask, perhaps, why this power is not used, if the statement be true, I have to ask in return, why is the power of steam so lately come to application? so many millions of men boiled water every day for many thousand years; they must have frequently seen that boiling water, in tightly closed pots or kettles, would lift the cover or burst the vessel with great violence. The power of steam was, therefore, as commonly known down to the least kitchen or wash-woman, as the power of wind; but close observation and reflection were bestowed neither on the one nor the other."

Men having discovered the power of falling water, which after all is comparatively slight, how eagerly do they seek out and improve these *privileges*? Let a difference of but a few feet in level be discovered on some stream near a populous town, some slight occasion for gravity to act, and the whole economy of the neighborhood is changed at once. Men do indeed speculate about and with this power as if it were the only privilege. But meanwhile this aerial stream is falling from far greater heights with more constant flow, never shrunk by drought, offering mill-sites wherever the wind blows; a Niagara in the air, with no Canada side;—only the application is hard.

There are the powers too of the Tide and Waves, constantly ebbing and flowing, lapsing and relapsing, but they serve man in but few ways. They turn a few tide mills, and perform a few other insignificant and accidental services only. We all perceive the effect of the tide; how imperceptibly it creeps up into our harbors and rivers, and raises the heaviest navies as easily as the lightest ship. Everything that floats must yield to it. But man, slow to take nature's constant hint of assistance, makes slight and irregular use of this power, in careening ships and getting them afloat when aground.

The following is Mr. Etzler's calculation on this head: To form a conception of the power which the tide affords, let us imagine a surface of 100 miles square, or 10,000 square miles, where the tide rises and sinks, on an average, 10 feet; how many men would it require to empty a basin of 10,000 square miles area, and 10 feet deep, filled with sea-water, in 6¼ hours and fill it again in the same time? As one man can raise 8 cubic feet of seawater per minute, and in 6¼ hours 3,000, it would take 1,200,000,000 men, or as they could work only half the time, 2,400,000,000, to raise 3,000,000,000,000 cubic feet, or the whole quantity required in the given time.

This power may be applied in various ways. A large body, of the heaviest materials that will float, may first be raised by it, and being attached to the end of a balance reaching from the land, or from a stationary support, fastened to the bottom, when the tide falls, the whole weight will be brought to bear upon the end of the balance. Also when the tide rises it may be made to exert a nearly equal force in the opposite direction. It can be employed whenever a *point d'appui* can be obtained.

> "However, the application of the tide being by establishments fixed on the ground, it is natural to begin with them near the shores in shallow water, and upon sands, which may be extended gradually further into the sea. The shores of the continent, islands, and sands, being generally surrounded by shallow water, not exceeding from 50 to 100 fathoms in depth, for 20, 50, or 100 miles and upward. The coasts of North America, with their extensive sand-banks, islands, and rocks, may easily afford, for this purpose, a ground about 3,000 miles long, and, on an average, 100 miles broad, or 300,000 square miles, which, with a power of 240,000 men per square mile, as stated, at 10 feet tide, will be equal to 72,000 millions of men, or for every mile of coast, a power of 24,000,000 men.
>
> "Rafts, of any extent, fastened on the ground of the sea, along the shore, and stretching far into the sea, may be covered with fertile soil, bearing vegetables and trees, of every description, the finest gardens, equal to those the firm land may admit of, and buildings and machineries, which may operate, not only on the sea, where they are, but which also, by means of mechanical connections, may extend their operations for many miles into the continent. (Etzler's Mechanical System,

page 24.) Thus this power may cultivate the artificial soil for many miles upon the surface of the sea, near the shores, and, for several miles, the dry land, along the shore, in the most superior manner imaginable; it may build cities along the shore, consisting of the most magnificent palaces, every one surrounded by gardens and the most delightful sceneries; it may level the hills and unevennesses, or raise eminences for enjoying open prospect into the country and upon the sea; it may cover the barren shore with fertile soil, and beautify the same in various ways; it may clear the sea of shallows, and make easy the approach to the land, not merely of vessels, but of large floating islands, which may come from, and go to distant parts of the world, islands that have every commodity and security for their inhabitants which the firm land affords."

"Thus may a power, derived from the gravity of the moon and the ocean, hitherto but the objects of idle curiosity to the studious man, be made eminently subservient for creating the most delightful abodes along the coasts, where men may enjoy at the same time all the advantages of sea and dry land; the coasts may hereafter be continuous paradisiacal skirts between land and sea, everywhere crowded with the densest population. The shores and the sea along them will be no more as raw nature presents them now, but everywhere of easy and charming access, not even molested by the roar of waves, shaped as it may suit the purposes of their inhabitants; the sea will be cleared of every obstruction to free passage everywhere, and its productions in fishes, etc., will be gathered in large, appropriate receptacles, to present them to the inhabitants of the shores and of the sea."

Verily, the land would wear a busy aspect at the spring and neap tide, and these island ships—these *terræ infirmæ*—which realise the fables of antiquity, affect our imagination. We have often thought that the fittest locality for a human dwelling was on the edge of the land, that there the constant lesson and impression of the sea might sink deep into the life and character of the landsman, and perhaps impart a marine tint to his imagination. It is a noble word, that *mariner*—one who is conversant with the sea. There should be more of what it signifies in each of us. It is a worthy country to belong to—we look to see him not disgrace it. Perhaps we should be equally mariners and terreners, and even our Green Mountains need some of that sea-green to be mixed with them.

The computation of the power of the waves is less satisfactory. While only the average power of the wind, and the average height of the tide, were taken before now, the extreme height of the waves is used, for they are made to rise ten feet above the level of the sea, to which, adding ten more for depression, we have twenty feet, or the extreme height of a wave. Indeed, the power of the waves, which is produced by the wind blowing obliquely and at disadvantage upon the water, is made to be, not only three thousand times greater than that of the tide, but one hundred times greater than that of the wind itself, meeting its object at right angles. Moreover, this power is measured by the area of the vessel, and not by its length mainly, and it seems to be forgotten that the motion of the waves is chiefly undulatory, and exerts a power only within the limits of a vibration, else the very continents, with their extensive coasts, would soon be set adrift.

Finally, there is the power to be derived from Sunshine, by the principle on which Archimedes contrived his burning mirrors, a multiplication of mirrors reflecting the rays of the sun upon the same spot, till the requisite degree of heat is obtained. The principal application of this power will be to the boiling of water and production of steam.

> "How to create rivulets of sweet and wholesome water, on floating islands, in the midst of the ocean, will be no riddle now. Sea-water changed into steam, will distil into sweet water, leaving the salt on the bottom. Thus the steam engines on floating islands, for their propulsion and other mechanical purposes, will serve, at the same time, for the distillery of sweet water, which, collected in basins, may be led through channels over the island, while, where required, it may be refrigerated by artificial means, and changed into cool water, surpassing, in salubrity, the best spring water, because nature hardly ever distils water so purely, and without admixture of less wholesome matter."

So much for these few and more obvious powers, already used to a trifling extent. But there are innumerable others in nature, not described nor discovered. These, however, will do for the present. This would be to make the sun and the moon equally our satellites. For, as the moon is the cause of the tides, and the sun the cause of the wind, which, in turn, is the cause of the waves, all the work of this planet would be performed by these far influences.

"But as these powers are very irregular and subject to interruptions; the next object is to show how they may be converted into powers that operate continually and uniformly for ever, until the machinery be worn out, or, in other words, into perpetual motions." . . . "Hitherto the power of the wind has been applied immediately upon the machinery for use, and we have had to wait the chances of the wind's blowing; while the operation was stopped as soon as the wind ceased to blow. But the manner, which I shall state hereafter, of applying this power, is to make it operate only for collecting or storing up power, and then to take out of this store, at any time, as much as may be wanted for final operation upon the machines. The power stored up is to react as required, and may do so long after the original power of the wind has ceased. And though the wind should cease for intervals of many months, we may have by the same power a uniform perpetual motion in a very simple way."

"The weight of a clock being wound up gives us an image of reaction. The sinking of this weight is the reaction of winding it up. It is not necessary to wait till it has run down before we wind up the weight, but it may be wound up at any time, partly or totally; and if done always before the weight reaches the bottom, the clock will be going perpetually. In a similar, though not in the same way, we may cause a reaction on a larger scale. We may raise, for instance, water by the immediate application of wind or steam to a pond upon some eminence, out of which, through an outlet, it may fall upon some wheel or other contrivance for setting machinery a going. Thus we may store up water in some eminent pond, and take out of this store, at any time, as much water through the outlet as we want to employ, by which means the original power may react for many days after it has ceased." . . . "Such reservoirs of moderate elevation or size need not be made artificially, but will be found made by nature very frequently, requiring but little aid for their completion. They require no regularity of form. Any valley with lower grounds in its vicinity, would answer the purpose. Small crevices may be filled up. Such places may be eligible for the beginning of enterprises of this kind."

The greater the height, of course the less water required. But suppose a level and dry country; then hill and valley, and "eminent pond," are to be constructed by main force; or if the springs are unusually low, then dirt and stones may be used, and the disadvantage arising from friction will be counterbalanced by their greater gravity. Nor shall a single rood of dry land be sunk in such artificial ponds as may be wasted, but their surfaces "may be covered with rafts decked with fertile earth, and all kinds of vegetables which may grow there as well as anywhere else."

And finally, by the use of thick envelopes retaining the heat, and other contrivances, "the power of steam caused by sunshine may react at will, and thus be rendered perpetual, no matter how often or how long the sunshine may be interrupted. (Etzler's Mechanical System)."

Here is power enough, one would think, to accomplish somewhat. These are the powers below. Oh ye millwrights, ye engineers, ye operatives and speculators of every class, never again complain of a want of power; it is the grossest form of infidelity. The question is not how we shall execute, but what. Let us not use in a niggardly manner what is thus generously offered.

Consider what revolutions are to be effected in agriculture. First, in the new country, a machine is to move along taking out trees and stones to any required depth, and piling them up in convenient heaps; then the same machine, "with a little alteration," is to plane the ground perfectly, till there shall be no hills nor valleys, making the requisite canals, ditches and roads, as it goes along. The same machine, "with some other little alterations," is then to sift the ground thoroughly, supply fertile soil from other places if wanted, and plant it; and finally, the same machine "with a little addition," is to reap and gather in the crop, thresh and grind it, or press it to oil, or prepare it any way for final use. For the description of these machines we are referred to "Etzler's Mechanical System, page 11 to 27." We should be pleased to see that "Mechanical System," though we have not been able to ascertain whether it has been published, or only exists as yet in the design of the author. We have great faith in it. But we cannot stop for applications now.

> "Any wilderness, even the most hideous and sterile, may be converted into the most fertile and delightful gardens. The most dismal swamps may be cleared of all their spontaneous growth, filled up and levelled, and intersected by canals, ditches and aqueducts, for draining them entirely. The soil, if required, may be meliorated, by covering or mixing it with rich soil taken from distant places, and the same be mouldered to fine dust, levelled, sifted from all roots, weeds and stones, and sowed and

> planted in the most beautiful order and symmetry, with fruit trees and vegetables of every kind that may stand the climate."

New facilities for transportation and locomotion are to be adopted:

> "Large and commodious vehicles, for carrying many thousand tons, running over peculiarly adapted level roads, at the rate of forty miles per hour, or one thousand miles per day, may transport men and things, small houses, and whatever may serve for comfort and ease, by land. Floating islands, constructed of logs, or of wooden-stuff prepared in a similar manner, as is to be done with stone, and of live trees, which may be reared so as to interlace one another, and strengthen the whole, may be covered with gardens and palaces, and propelled by powerful engines, so as to run at an equal rate through seas and oceans. Thus, man may move, with the celerity of a bird's flight, in terrestrial paradises, from one climate to another, and see the world in all its variety, exchanging, with distant nations, the surplus of productions. The journey from one pole to another may be performed in a fortnight; the visit to a transmarine country in a week or two; or a journey round the world in one or two months by land and water. And why pass a dreary winter every year while there is yet room enough on the globe where nature is blessed with a perpetual summer, and with a far greater variety and luxuriance of vegetation? More than one-half of the surface of the globe has no winter. Men will have it in their power to remove and prevent all bad influences of climate, and to enjoy, perpetually, only that temperature which suits their constitution and feeling best."

Who knows but by accumulating the power until the end of the present century, using meanwhile only the smallest allowance, reserving all that blows, all that shines, all that ebbs and flows, all that dashes, we may have got such a reserved accumulated power as to run the earth off its track into a new orbit, some summer, and so change the tedious vicissitude of the seasons? Or, perchance, coming generations will not abide the dissolution of the globe, but, availing themselves of future inventions in aerial locomotion, and the navigation of space, the entire race may migrate from the earth, to settle some vacant and more western planet, it may be still healthy, perchance unearthy, not composed of dirt and stones, whose primary strata only are strewn, and where no weeds are sown. It took but little art, a simple application of natural laws, a canoe, a paddle, and a sail of matting, to people the isles of the Pacific, and a little more will people the shining isles of space. Do we not see in the firmament the lights carried along the shore by night, as Columbus did? Let us not despair nor mutiny.

> "The dwellings also ought to be very different from what is known, if the full benefit of our means is to [be] enjoyed. They are to be of a structure for which we have no name yet. They are to be neither palaces, nor temples, nor cities, but a combination of all, superior to whatever is known. Earth may be baked into bricks, or even vitrified stone by heat,—we may bake large masses of any size and form into stone and vitrified substance of the greatest durability, lasting even thousands of years, out of clayey earth, or of stones ground to dust, by the application of burning mirrors. This is to be done in the open air, without other preparation than gathering the substance, grinding and mixing it with water and cement, moulding or casting it, and bringing the focus of the burning mirrors of proper size upon the same. The character of the architecture is to be quite different from what it ever has been hitherto; large solid masses are to be baked or cast in one piece, ready shaped in any form that may be desired. The building may, therefore, consist of columns two hundred feet high and upwards, of proportionate thickness, and of one entire piece of vitrified substance; huge pieces are to be moulded so as to join and hook on to each other firmly, by proper joints and folds, and not to yield in any way without breaking.
>
> "Foundries, of any description, are to be heated by burning mirrors, and will require no labor, except the making of the first moulds and the superintendence for gathering the metal and taking the finished articles away."

Alas, in the present state of science, we must take the finished articles away; but think not that man will always be a victim of circumstances.

The countryman who visited the city and found the streets cluttered with bricks and lumber, reported that it was not yet finished, and one who considers the endless repairs and reforming of our houses, might well wonder when they will be done. But why may not the dwellings of men on this earth be built once for all of some durable material, some Roman or Etruscan masonry which will stand, so that time shall only adorn and beautify them? Why may we not finish the outward world for posterity, and leave them leisure to attend to the inner? Surely, all the gross necessities and economies might be cared for in a few years. All might be built and baked and stored up, during this, the term-time of the world, against the vacant eternity, and the globe go provisioned and furnished like our public vessels, for its voyage through space, as through some Pacific ocean, while we would "tie up the rudder and sleep before the wind," as those who sail from Lima to Manilla.

But, to go back a few years in imagination, think not that life in these crystal palaces is to bear any analogy to life in our present humble cottages. Far from it. Clothed, once for all, in some "flexible stuff," more durable than George Fox's suit of leather, composed of "fibres of vegetables," "glutinated" together by some "cohesive substances," and made into sheets, like paper, of any size or form, man will put far from him corroding care and the whole host of ills.

> "The twenty-five halls in the inside of the square are to be each two hundred feet square and high; the forty corridors, each one hundred feet long and twenty wide; the eighty galleries, each from 1,000 to 1,250 feet long; about 7,000 private rooms, the whole surrounded and intersected by the grandest and most splendid colonnades imaginable; floors, ceilings, columns with their various beautiful and fanciful intervals, all shining, and reflecting to infinity all objects and persons, with splendid lustre of all beautiful colors, and fanciful shapes and pictures. All galleries, outside and within the halls, are to be provided with many thousand commodious and most elegant vehicles, in which persons may move up and down, like birds, in perfect security, and without exertion. Any member may procure himself all the common articles of his daily wants, by a short turn of some crank, without leaving his apartment; he may, at any time, bathe himself in cold or warm water, or in steam, or in some artificially prepared liquor for invigorating health. He may, at any time, give to the air in his apartment that temperature that suits his feeling best. He may cause, at any time, an agreeable scent of various kinds. He may, at any time, meliorate his breathing air,—that main vehicle of vital power. Thus, by a proper application of the physical knowledge of our days, man may be kept in a perpetual serenity of mind, and if there is no incurable disease or defect in his organism, in constant vigor of health, and his life be prolonged beyond any parallel which present times afford.
>
> "One or two persons are sufficient to direct the kitchen business. They have nothing else to do but to superintend the

cookery, and to watch the time of the victuals being done, and then to remove them, with the table and vessels, into the dining-hall, or to the respective private apartments, by a slight motion of the hand at some crank. Any extraordinary desire of any person may be satisfied by going to the place where the thing is to be had; and anything that requires a particular preparation in cooking or baking, may be done by the person who desires it."

This is one of those instances in which the individual genius is found to consent, as indeed it always does, at last, with the universal. These last sentences have a certain sad and sober truth, which reminds us of the scripture of all nations. All expression of truth does at length take the deep ethical form. Here is hint of a place the most eligible of any in space, and of a servitor, in comparison with whom, all other helps dwindle into insignificance. We hope to hear more of him anon, for even crystal palace would be deficient without his invaluable services.

And as for the environs of the establishment,

"There will be afforded the most enrapturing views to be fancied, out of the private apartments, from the galleries, from the roof, from its turrets and cupolas,—gardens as far as the eye can see, full of fruits and flowers, arranged in the most beautiful order, with walks, colonnades, aqueducts, canals, ponds, plains, amphitheatres, terraces, fountains, sculptural works, pavilions, gondolas, places for public amusement, etc., to delight the eye and fancy, the taste and smell." . . . "The walks and roads are to be paved with hard vitrified, large plates, so as to be always clean from all dirt in any weather or season. . . . The channels being of vitrified substance, and the water perfectly clear, and filtrated or distilled if required, may afford the most beautiful scenes imaginable, while a variety of fishes is seen clear down to the bottom playing about, and the canals may afford at the same time, the means of gliding smoothly along between various sceneries of art and nature, in beautiful gondolas, while their surface and borders may be covered with fine land and aquatic birds. The walks may be covered with porticos adorned with magnificent columns, statues and sculptural works; all of vitrified substance, and lasting for ever, while the beauties of nature around heighten the magnificence and deliciousness."

"The night affords no less delight to fancy and feelings. An infinite variety of grand, beautiful and fanciful objects and sceneries, radiating with crystalline brilliancy, by the illumination of gaslight; the human figures themselves, arrayed in the most beautiful pomp fancy may suggest, or the eye desire, shining even with brilliancy of stuffs and diamonds, like stones of various colors, elegantly shaped and arranged around the body; all reflected a thousand-fold in huge mirrors and reflectors of various forms; theatrical scenes of a grandeur and magnificence, and enrapturing illusions, unknown yet, in which any person may be either a spectator or actor; the speech and the songs reverberating with increased sound, rendered more sonorous and harmonious than by nature, by vaultings that are moveable into any shape at any time; the sweetest and most impressive harmony of music, produced by song and instruments partly not known yet, may thrill through the nerves and vary with other amusements and delights.

"At night the roof, and the inside and outside of the whole square, are illuminated by gas-light, which in the mazes of many-colored crystal-like colonnades and vaultings, is reflected with a brilliancy that gives to the whole a lustre of precious stones, as far as the eye can see,—such are the future abodes of men." . . . "Such is the life reserved to true intelligence, but withheld from ignorance, prejudice, and stupid adherence to custom." . . . "Such is the domestic life to be enjoyed by every human individual that will partake of it. Love and affection may there be fostered and enjoyed without any of the obstructions that oppose, diminish, and destroy them in the present state of men." . . . "It would be as ridiculous, then, to dispute and quarrel about the means of life, as it would be now about water to drink along mighty rivers, or about the permission to breathe air in the atmosphere, or about sticks in our extensive woods."

Thus is Paradise to be Regained, and that old and stern decree at length reversed. Man shall no more earn his living by the sweat of his brow. All labor shall be reduced to "a short turn of some crank," and "taking the finished article away." But there is a crank,—oh, how hard to be turned! Could there not be a crank upon a crank,—an infinitely small crank?—we would fain inquire. No,—alas! not. But there is a certain divine energy in every man, but sparingly employed as yet, which may be called the crank within,—the crank after all,—the prime mover in all machinery,—quite indispensable to all work. Would that we might get our hands on its handle! In fact no work can be shirked. It may be postponed indefinitely, but not infinitely. Nor can any really important work be made easier by co-operation or machinery. Not one particle of labor now threatening any man can be routed without being performed. It cannot be hunted out of the vicinity like jackals and hyenas. It will not run. You may begin by sawing the little sticks, or you may saw the great sticks first, but sooner or later you must saw them both.

We will not be imposed upon by this vast application of forces. We believe that most things will have to be accomplished still by the application called Industry. We are rather pleased after all to consider the small private, but both constant and accumulated force, which stands behind every spade in the field. This it is that makes the valleys shine, and the deserts really bloom. Sometimes, we confess, we are so degenerate as to reflect with pleasure on the days when men were yoked like cattle, and drew a crooked stick for a plough. After all, the great interests and methods were the same.

It is a rather serious objection to Mr. Etzler's schemes, that they require time, men, and money, three very superfluous and inconvenient things for an honest and well-disposed man to deal with. "The whole world," he tells us, "might therefore be really changed into a paradise, within less than ten years, commencing from the first year of an association for the purpose of constructing and applying the machinery." We are sensible of a startling incongruity when time and money are mentioned in this connection. The ten years which are proposed would be a tedious while to wait, if every man were at his post and did his duty, but quite too short a period, if we are to take time for it. But this fault is by no means peculiar to Mr. Etzler's schemes. There is far too much hurry and bustle, and too little patience and privacy, in all our methods, as if something were to be accomplished in centuries. The true reformer does not want time, nor money, nor co-operation, nor advice. What is time but the stuff delay is made of? And depend upon it, our virtue will not live on the interest of our money. He expects no income but our outgoes; so soon as we begin to count the cost the cost begins. And as for advice, the information floating in the atmosphere of society is as evanescent and unserviceable to him as gossamer for clubs of Hercules. There is absolutely no common sense; it is common nonsense. If we are to risk a cent or a drop of our blood, who then shall advise us? For ourselves, we are too young for experience. Who is old enough? We are older by faith than by experience. In the unbending of the arm to do the deed there is experience worth all the maxims in the world.

> "It will now be plainly seen that the execution of the proposals is not proper for individuals. Whether it be proper for government at this time, before the subject has become popular, is a question to be decided; all that is to be done, is to step forth, after mature reflection, to confess loudly one's conviction, and to constitute societies. Man is powerful but in union with many. Nothing great, for the improvement of his own condition, or that of his fellow men, can ever be effected by individual enterprise."

Alas! this is the crying sin of the age, this want of faith in the prevalence of a man. Nothing can be effected but by one man. He who wants help wants everything. True, this is the condition of our weakness, but it can never be the means of our recovery. We must first succeed alone, that we may enjoy our success together. We trust that the social movements which we witness indicate an aspiration not to be thus cheaply satisfied. In this matter of reforming the world, we have little faith in corporations; not thus was it first formed.

But our author is wise enough to say, that the raw materials for the accomplishment of his purposes, are "iron, copper, wood, earth chiefly, and a union of men whose eyes and understanding are not shut up by preconceptions." Aye, this last may be what we want mainly,—a company of "odd fellows" indeed.

"Small shares of twenty dollars will be sufficient,"—in all, from "200,000 to 300,000,"—"to create the first establishment for a whole community of from 3000 to 4000 individuals"—at the end of five years we shall have a principal of 200 millions of dollars, and so paradise will be wholly regained at the end of the tenth year. But, alas, the ten years have already elapsed, and there are no signs of Eden yet, for want of the requisite funds to begin the enterprise in a hopeful manner. Yet it seems a safe investment. Perchance they could be hired at a low rate, the property being mortgaged for security, and, if necessary, it could be given up in any stage of the enterprise, without loss, with the fixtures.

Mr. Etzler considers this "Address as a touchstone, to try

whether our nation is in any way accessible to these great truths, for raising the human creature to a superior state of existence, in accordance with the knowledge and the spirit of the most cultivated minds of the present time." He has prepared a constitution, short and concise, consisting of twenty-one articles, so that wherever an association may spring up, it may go into operation without delay; and the editor informs us that "Communications on the subject of this book may be addressed to C. F. Stollmeyer, No. 6, Upper Charles street, Northampton square, London."

But we see two main difficulties in the way. First, the successful application of the powers by machinery, (we have not yet seen the "Mechanical System,"), and, secondly, which is infinitely harder, the application of man to the work by faith. This it is, we fear, which will prolong the ten years to ten thousand at least. It will take a power more than "80,000 times greater than all the men on earth could effect with their nerves," to persuade men to use that which is already offered them. Even a greater than this physical power must be brought to bear upon the moral power. Faith, indeed, is all the reform that is needed; it is itself a reform. Doubtless, we are as slow to conceive of Paradise as of Heaven, of a perfect natural as of a perfect spiritual world. We see how past ages have loitered and erred; "Is perhaps our generation free from irrationality and error? Have we perhaps reached now the summit of human wisdom, and need no more to look out for mental or physical improvement?" Undoubtedly, we are never so visionary as to be prepared for what the next hour may bring forth.

Μέλλει τὸ θεῖον δ ἔστι τοιοῦτον φυσει.

The Divine is about to be, and such is its nature. In our wisest moments we are secreting a matter, which, like the lime of the shell fish, incrusts us quite over, and well for us, if, like it, we cast our shells from time to time, though they be pearl and of fairest tint. Let us consider under what disadvantages science has hitherto labored before we pronounce thus confidently on her progress.

> "There was never any system in the productions of human labor; but they came into existence and fashion as chance directed men." "Only a few professional men of learning occupy themselves with teaching natural philosophy, chemistry, and

> the other branches of the sciences of nature, to a very limited extent, for very limited purposes, with very limited means." "The science of mechanics is but in a state of infancy. It is true, improvements are made upon improvements, instigated by patents of government; but they are made accidentally or at hap-hazard. There is no general system of this science, mathematical as it is, which developes its principles in their full extent, and the outlines of the application to which they lead. There is no idea of comparison between what is explored and what is yet to be explored in this science. The ancient Greeks placed mathematics at the head of their education. But we are glad to have filled our memory with notions, without troubling ourselves much with reasoning about them."

Mr. Etzler is not one of the enlightened practical men, the pioneers of the actual, who move with the slow deliberate tread of science, conserving the world; who execute the dreams of the last century, though they have no dreams of their own; yet he deals in the very raw but still solid material of all inventions. He has more of the practical than usually belongs to so bold a schemer, so resolute a dreamer. Yet his success is in theory, and not in practice, and he feeds our faith rather than contents our understanding. His book wants order, serenity, dignity, everything—but it does not fail to impart what only man can impart to man of much importance, his own faith. It is true his dreams are not thrilling nor bright enough, and he leaves off to dream where he who dreams just before the dawn begins. His castles in the air fall to the ground, because they are not built lofty enough; they should be secured to heaven's roof. After all, the theories and speculations of men concern us more than their puny execution. It is with a certain coldness and languor that we loiter about the actual and so called practical. How little do the most wonderful inventions of modern times detain us. They insult nature. Every machine, or particular application, seems a slight outrage against universal laws. How many fine inventions are there which do not clutter the ground? We think that those only succeed which minister to our sensible and animal wants, which bake or brew, wash or warm, or the like. But are those of no account which are patented by fancy and imagination, and succeed so admirably in our dreams that they give the tone still to our waking thoughts? Already nature is serving all those uses which science slowly derives on a much higher and grander scale to him that will be served by her. When the sunshine falls on the path of the poet, he enjoys all those pure benefits and pleasures which the arts slowly and partially realize from age to age. The winds which fan his cheek waft him the sum of that profit and happiness which their lagging inventions supply.

The chief fault of this book is, that it aims to secure the greatest degree of gross comfort and pleasure merely. It paints a Mahometan's heaven, and stops short with singular abruptness when we think it is drawing near to the precints of the Christian's,—and we trust we have not made here a distinction without a difference. Undoubtedly if we were to reform this outward life truly and thoroughly, we should find no duty of the inner omitted. It would be employment for our whole nature; and what we should do thereafter would be as vain a question as to ask the bird what it will do when its nest is built and its brood reared. But a moral reform must take place first, and then the necessity of the other will be superseded, and we shall sail and plough by its force alone. There is a speedier way than the Mechanical System can show to fill up marshes, to drown the roar of the waves, to tame hyænas, secure agreeable environs, diversify the land, and refresh it with "rivulets of sweet water," and that is by the power of rectitude and true behavior. It is only for a little while, only occasionally, methinks, that we want a garden. Surely a good man need not be at the labor to level a hill for the sake of a prospect, or raise fruits and flowers, and construct floating islands, for the sake of a paradise. He enjoys better prospects than lie behind any hill. Where an angel travels it will be paradise all the way, but where Satan travels it will be burning marl and cinders. What says Veeshnoo Sunma? "He whose mind is at ease is possessed of all riches. Is it not the same to one whose foot is enclosed in a shoe, as if the whole surface of the earth were covered with leather?"

He who is conversant with the supernal powers will not worship these inferior deities of the wind, the waves, tide, and sunshine. But we would not disparage the importance of such calculations as we have described. They are truths in physics, because they are true in ethics. The moral powers no one would presume to calculate. Suppose we could compare the moral with the physical, and say how many horse-power the force of love, for instance, blowing on every square foot of a man's soul, would equal. No doubt we are well aware of this force; figures would not increase our respect for it; the sunshine is equal to but one ray of its heat. The light of the sun is but the shadow of love. "The souls of men loving and fearing God," says Raleigh, "receive influence from that divine light itself, whereof the sun's elasity, and that of the stars, is by Plato called but a shadow. *Lumen est umbra Dei, Deus est Lumen Luminis.* Light is the shadow of God's brightness, who is "the light of light," and, we may add, the heat of heat. Love is the wind, the tide, the waves, the sunshine. Its power is incalculable; it is many horse power. It never ceases, it never slacks; it can move the globe without a resting-place; it can warm without fire; it can feed without meat; it can clothe without garments; it can shelter without roof; it can make a paradise within which will dispense with a paradise without. But though the wisest men in all ages have labored to publish this force, and every human heart is, sooner or later, more or less, made to feel it, yet how little is actually applied to social ends. True, it is the motive power of all successful social machinery; but, as in physics, we have made the elements do only a little drudgery for us, steam to take the place of a few horses, wind of a few oars, water of a few cranks and hand-mills; as the mechanical forces have not yet been generously and largely applied to make the physical world answer to the ideal, so the power of love has been but meanly and sparingly applied, as yet. It has patented only such machines as the almshouses, the hospital, and the Bible Society, while its infinite wind is still blowing, and blowing down these very structures, too, from time to time. Still less are we accumulating its power, and preparing to act with greater energy at a future time. Shall we not contribute our shares to this enterprise, then?

INTRODUCTORY NOTE

If technology can be defined as man making a material and useful environment, then the title of Ewbank's book, *The World a Workshop,* captures the essence of technological activity. Ewbank insisted that God created the earth—"this spherical mass of materials"—so that man could manipulate its forces and its inert, unshaped, and unwrought materials to make it a habitable artifice. (A century later, social critics and environmentalists would lament man's handiwork).

Ewbank had no patience with those who viewed man as a relatively passive tiller of the soil and tender of the animals; the earth was a workshop, a manufactory, not a garden or a farm. The earth was no "caravanserai" to be left unworked. He called to mind the pleasant American home with garden in front and orchard and cornfield behind, but quickly shifted attention with, "but, hark, a train or car is approaching," and demanded that his reader focus upon the technological realities sustaining the American way of life: "potteries, tanneries, grist, saw, paper, and cotton mills; foundries; machine shops. . . ." Man the maker—technological man—Ewbank insisted, had to imitate "the Great Artificer of all that moves"; man's mind must not passively contemplate the world around it, but should act upon it.

Thomas Ewbank (1792–1870) served as U.S. Commissioner of Patents from 1849 to 1852. Born in Durham, England, he emigrated in 1819 to New York. He had been an apprentice tin- and coppersmith, and he read deeply and widely in the industrial arts, science, and engineering. In the United States he was successful in metal manufacturing, but after 1836 he devoted himself entirely to study, travel, and writing. His books include technical studies and popular essays on history, invention, and applied sciences. As commissioner of patents, he published his wide-ranging and speculative essays in the Patent Office reports.

The World a Workshop

THOMAS EWBANK

There is a series of initial and very remarkable facts in the present condition and disposition of the matter of our globe that have, I think, been little regarded. They have not, that I am aware, been considered in the light in which they are here presented to the reader, although singularly illustrative of the character here given to man—a character with which all terrestrial phenomena accord; the volume of the Earth, the quantity, varieties, and qualities of its materials; the difference between its internal and external products, its central fires and internal movements, its atmospheric envelope, the alternate illuminating and darkening of its surface; the physical and mental constitution of man; his necessities, enjoyments, aspirations, and morals; the duration of his life, and his education for a more enlarged sphere of existence.

The title has been selected for the purpose of arresting some fugitive thoughts, in the hope that scientific writers will add to their number, amplify and arrange them into a regular treatise on a subject fertile as the earth itself; and under the impression that they will serve to explode those vague impressions that so generally prevail respecting the real relationship between us and the globe we inhabit. Instead of universal confidence arising from settled conviction, uncertainty is almost everywhere felt as regards the extent to which we should give up ourselves to the earth. On the one hand it is viewed as a mere caravanserai, a temporary convenience for passing travellers, and therefore unworthy of more than passing regard: on the other are men active in exploring its resources, drawing from it elements of progressive civilization, and reflecting streams of light on the attributes of the Creator, and on his intentions towards us.

Thomas Ewbank, *The World a Workshop: The Physical Relationship of Man to the Earth* (New York: Appleton, 1855), pp. 19–24, 122–123, 172–173.

Let any one whose mind the subject never entered think within himself: For what was this spherical mass of materials made? To what special, or partial, or general uses was man to put them? Or was he to do anything with them? The questions involve the character and designs of the Creator, and the very basis of human welfare; for how are we to work out our destiny if we know not what it is?

It does seem strange that men should never have definitely determined the predominating characteristics of a world they have so long occupied; that they should have fluttered and fumed and busied themselves, like insects, on little isolated hillocks, instead of taking a view of the entire establishment and blocking out a working plan of the whole.

A glance at the countless occupations of men; at the divisions and subdivisions of material and speculative labor; at the bustle of trades, schemings of the ambitious, and struggles and competitions of all; would almost lead one to conclude we had been thrown at random on the earth, and left to scramble by chance for a living upon it. Reflection, however, points to a rule or law prevading the hubbub, and shows the diversities of pursuits to be off-shoots of a primordial one which arises out of the constitution of the globe and our own organization.

It is conceded that the earth is in every respect fitted for man and he for it; and it might *à priori* be inferred that amid the multitudinous adaptations, there would be some ruling one to which the rest were to be collateral, or in which they centred. It was not probable that his abode was to consist of a mass of miscellaneous accommodations without character as a whole, or that his faculties were to be a heterogeneous compound without reference to some predominating task on which they were to be employed. What then was it that was so conspicuously to mark his connexion with the earth, and more than anything else proclaim him lord or lessee of it? It was the character he was to assume as a MANIPULATOR OF MATTER. The earth was to be a manufactory and he a manufacturer. It was to furnish him with unwrought material, while the sounds of his implements acting upon it were to swell till their reverberations rolled over the whole. His connexion with it, then, arises from the work assigned him, the materials for it, and the uses to which he is to put them.

It is, of course, from the materials of the earth and their

attributes that its character is to be deduced. If they are adapted to man's wants, and to be operated on by him; if they are indispensable to him and yet useless till manipulated; it must needs follow that it was designed for a Factory. If it were wholly vegetable, it would be a Farm; if its products were objects ready for use, a Bazaar. But almost the whole is mineral —inert, unshapen, and unwrought, while even vegetable and animal substances require elaboration.

If, as some say, the materials were collected and arranged as we find them, to afford a platform for living creatures and a surface vegetation to sustain them, the vast subterrene minerals would be superfluous—an idea utterly irreconcilable with the wisdom of their Author. The hypothesis, also, that the chief employment of man was to till the soil and raise cattle, is an unworthy one, since it puts us much on a par with the lower tribes, in making the procuring and consuming of food the principal object of our being. If we were made to live like cattle—merely to eat and sleep—it would be true, and the earth might then be considered a mere victualling institution. But with us, and all intelligences, food is, like the traveller's staff, an adjunct of life, a mere aid in accomplishing the purposes of existence. Hence, though agriculture must always be a marked and honored department of labor, it is not the only one, nor the chief one. The portion of the earth's materials taken up by it is moreover comparatively small, and therefore the bulk of mundane matter was designed for something to which the raising of food was to be subservient.

While most persons think not, and care not, what the prominent character of the planet is, many view it in aspects congenial to themselves: a theatre for politicians, a battle-field for warriors, a court for lawyers, a loungingplace for people of fashion and leisure, &c. Morbid minds make it a crowded hospital and charnel-house, while the Indian believes it was made for nothing else than hunting game in. With these, and all surface dreamers, its vast underground treasures are not thought of. The inorganic world, its forces, principles, and processes, are to them as if they were not. It is only as a Factory, a General Factory, that the whole materials and influences of the earth are to be brought into play; and with this professional character of our globe every feature in creation will be found to harmonize.

For what classes then chiefly was the world of inorganic matter provided? Observe that dwelling; it belongs to a family neither rich nor poor; neat, commodious, and attractive in itself; it has a garden in front, an orchard and corn-field behind. Mark the social enjoyments, intelligence, and contentment of its inmates; the abundance of necessaries, of comforts and conveniences; the ornaments and elegances in dress and furniture, with contributions from almost every productive and decorative art. But, hark! a train of cars is approaching. It stops one moment and starts the next with a shriek for the city, whirling us along level and undulating lands, through tunnelled mountains, over rivers on bridges of granite, and others of iron. In the quick-moving panorama arise before us, and in a moment pass by, brick and lime kilns; potteries; tanneries; grist, saw, paper, and cotton mills; foundries; machine shops; chair, cloth, and carpet factories. We come in sight of a bay, on which ships laden with foreign merchandise are floating in with the tide, and others with home manufactures passing out. Crossing over in a steamer we find an extensive border of leafless forest resolved into masts of vessels crowded into continuous docks, and on landing, feel the air rent and agitated, like rippled water, with the noise of stevedores and draymen. We have business to transact for a friend, and pick our way along the side-walks, among packing-cases of dry-goods, casks of hardware, bundles of sheet and hoop iron, and loads of other goods. Next we stop at a telegraph office, and in five minutes our friend, though two hundred miles distant, receives and answers our note. On leaving the street of merchants for others occupied by watchmakers, jewellers, opticians, philosophical and musical instrument makers, engravers and printers, we call at a newspaper office to insert an advertisement and order the daily sheet for a neighbor. Need we proceed? It was for men who bring such things out of inert matter that this world of matter was made.

. . .

If we revert to his [man's] entrance into the world, we shall be led to the conviction that he is created an artisan, and sent here to exercise his vocation. What were the things on which he opened his eyes, and which continued through life to arrest his attention, but those of a factory crowded with work!

"All things," saith the Preacher, "are full of labor," and so full that man cannot utter it. The more he observed the earth, the more he found it the theatre of ceaseless manufacturing activities:—Its forces, chemical, electrical, and mechanical, latent and palpable, imperceptible and overpowering, never sleep. Without intermission, they are converting elemental matter into animal, vegetable, and mineral, into solid, liquid, and gaseous bodies—ever breaking up and renewing them. Every foot of it is paved with influences that are decomposing ingredients of worn-out fabrics, preparatory to their being again made up into similar things, or in gathering them up for other commodities. There is no suspension of the work on account of the moving-machinery, no waste of power, and no refuse material—not a scrap but what is worked up. The supply keeps pace with the demand, and, as with human merchandise, the goods vary with what they are made of, and their style with changes of times and seasons:—a perfectly organized manufacturing establishment.

Born and brought up in the busy scene, confined for life to it, having no ideas but what are derived from it—a true factory child—what else could he do but in the theatre of action assigned him imitate

> "The Great Artificer of all that moves!"

His instincts, organization, and condition defined his profession, and compelled him to assume it. In no other capacity could he work his mission out. He saw that the Proprietor had furnished him with raw materials, not with finished goods. He had corn, but no bread; fuel, not furnaces; clay, but neither bricks nor tiles; sand, but not glass; textile substances were supplied, but he was to convert them into plain and ornamental fabrics: iron was given, but not in bars. The properties of alloys he was to find out, and discover the means of producing a prime material for cutting implements and tools. He essayed these things, and succeeding, his natural career opened before him.

. . .

Hence there is nothing here to encourage the absorption of life in spiritual thought. The mind is to be active, and active on something else than itself and for itself. Those who have no sympathy with things here, have no business here. It is in vain to create and decorate worlds for them. To float in some dreary corner of the abyss, out of sight of any orb, would answer their purposes better, since no object would then be near to break their meditations. The idea of this sphere being an inn to give a few nights' lodgings to travellers passing hastily over it, the object of whose journeying is too high and urgent to allow them to examine or care for its conveniences—whence could it have proceeded, but from the weakness that leads men to magnify their vocations, and infer thence their own importance? Such are the absurdities into which those fall who attempt to separate what God hath everywhere united—matter and mind; who insanely dream of pleasing him, by debasing one and exalting the other.

Man is to be no spectral recluse, dozing away life over legends and relics; soddening his soul to the condition of a mollusk, and limiting his views of the world, like those of an oyster, to his cell; leaving the earth to grow up a jungle, and his thews and sinews to wither for want of employment. No! He is to be no dreamer in-doors, but an active, vigorous, and refreshing thinker and worker without.

. . .

INTRODUCTORY NOTE

"Effects of Machinery" is a review of a book first published in England. The object of the book was to persuade the British workingman that it was counter to his best interests to oppose the introduction of the technology and industrial organization responsible for the Industrial Revolution. The intent of the American reviewer was to convince the reader that the argument in support of mechanization and industrialization could be made even more strongly for America. This country, the reviewer believed, provided a superb testing ground for the influence of the machine upon society, and he anticipated that that influence would be benevolent. America provided a superior area for the working out of the machine's destiny because the American environment was not burdened by a poor system of government, grinding taxation, and unequal laws, as was the case in England. The unhappy effects seeming to stem from England's industrialization in fact resulted primarily—the reviewer argued—from the political and economic system. And even these, the reviewer was confident, were but temporary dislocations caused by changing technology. In America, by contrast, the political system and the spirit of society precluded such unfortunate consequences (witness, the writer argued, the beautiful villages, the schools, the churches of industrial America).

The writer's enthusiasm for machinery surfaced as he answered its critics argument by argument. His utilitarian philosophy, his epithets for his age ("age of improvement," "age of practical benevolence"), his conviction that Americans were creating with technology the highest civilization from a wilderness, his anticipation that progress would continue to accelerate as it had begun to in the past fifty years were typical of the ebullient American attitude toward technology in the early nineteenth century.

Effects of Machinery [a review of] *The Working Man's Companion, No. 1*

The Results of Machinery, being an Address to the Working Men of the United Kingdom. pp. 216 American Edition. Philadelphia. 1831.

This little book was published under the superintendence of the Society for the Diffusion of Useful Knowledge. Its object is to convince the working men of the United Kingdom, of the folly and wickedness of attempting to arrest the progress of improvement, by the destruction of machinery. It is written in a plain, unadorned style, but it is replete with valuable facts, and strong and persuasive reasoning. We commend the book to all croakers,—to all praisers of the past and revilers of the present time. We ask a careful perusal of it, of those venerable grandmothers who see misery and ruin close at hand, because the sound of the spinning-wheel and the loom is no longer heard in all our farm-houses.

In the present article, we shall attempt an independent and somewhat enlarged discussion of the principal questions, presented in this work. We shall not confine our attention to 'cheap production and increased employment,' alone; but shall endeavor to trace the influence of machinery farther, in its effects on society. The question is, is this influence,—confessedly, and beyond calculation, vast,—good or evil? This has been said to be 'a far more difficult and complex question, than any that political economists have yet engaged with.' Its importance and interests are certainly not exceeded by its difficulty and complexity. The first arise from the intimate connexion of the influence of machinery with every other influence, that affects the social condition of man, and the last are shared by it in common with every other question, relating to matters yet imperfectly understood. It is embarrassed by being complicated with a number of considerations, not necessarily belonging to it, and because it requires, in those who would re-

North American Review, 34 (January 1832), 220–246.

solve it, a larger amount of contemporary information, than is generally, or can be easily, acquired.

We look upon the knowledge of the present circumstances of society, of the transactions of our own age and country, of modern science and modern art, as more important than any other. Yet it is precisely the sort of knowledge, of which, until very recently, we have had least. We would not be understood to undervalue any species of knowledge. Every kind of information is precious. We would only say, that that which instructs us, where we are, what we are, and how we are, has peculiar value. It is true, that to know the present, we must be, in some degree, acquainted with the past. To understand the result, we must have knowledge of the cause. To foresee consequences, it is necessary to know how consequences have been heretofore produced. What we complain of is, not that we know too much of what has been, but that we do not know enough of what is; —not that we are too familiar with the past, but that we are not familiar enough with the present. And we would go so far as to say, that, if any part of knowledge were to be given up, it would be better to let alone the study of what happened before we were born, and the conjecture of what is to happen after we are dead, and confine our view within the horizon of our present existence. It was demanded of the Spartan king, 'what study is fittest for the boy?' His answer was, 'that of the science most useful to the man.' Utility measures the value of knowledge, as of every thing else; and surely, on the scale of utility, the knowledge of what is all around us, affecting us, physically, intellectually, and morally, in countless ways, ranks far higher than the knowledge of the circumstances of preceding generations.

It has not been the fact, however, that men have applied themselves to the study of their own times, with as much earnestness, as to the investigation of the records of the past. It has always been extremely difficult to obtain contemporary information of events. Intelligence has been transmitted from point to point very slowly. And when it has finally reached its destination, it has come in so questionable a shape, that its authenticity could by no means be relied on. The consequence has been, that men of learning and study have turned away from so unpromising a field of research. Almost all writers, except those whose business was politics, have occupied themselves in other tasks. It was a natural consequence, that science became speculative, rather than practical. The object of study was rather to gratify the instinctive desire of knowledge, than to strike out a light to guide the conduct, or to discover the means of improving the condition of man. And thus men, instead of believing that they were intrusted by Providence with the care of their own fates, have been accustomed to think of themselves, as embarked, without a rudder, without a sail, without an oar, upon the stream of destiny, hurried on, they know not how, and destined to arrive, they know not whither.

But there is a better philosophy than this,—a philosophy that attributes more to men and less to circumstance. It teaches that knowledge is for use, and not for ostentation. It teaches that the great events, which crowd the historian's page, are beacons kindled by those who have gone before us, illumining the scene of present things, and dispelling, partially at least, the shadows, clouds and darkness, that overhang the future. Intelligence of recent events is now communicated with a degree of certainty and rapidity, utterly unknown hitherto. The best intellects are employed in the observation of the passing pageant of existence. The importance of each occurrence is immediately ascertained; its proper place in the system of events is fixed, and the fact, with the reasoning that links it to the past and the future, is communicated to the public, through the periodical press. It is true that the fact is yet frequently mis-stated, and the reasoning about it often erroneous; but, on the whole, truth greatly prevails; and the present age is doubtless better acquainted with itself, than any which have preceded it.

Still, this acquaintance, this self-knowledge, an attainment, by the way, quite as important to nations as to individuals, is extremely imperfect and superficial. A reflecting man who looks around him, upon the countless agencies, operating with different degrees of energy, for good or for evil, upon the condition and character of men, wherever man exists, cannot help feeling how little is known of things as they are. What we hear about the age in which we live, is quite too vague and general, to satisfy a rational curiosity. We hear it called the age of improvement, the enlightened age, the age of practical benevolence. But we want a deeper and more extensive knowledge, than these epithets convey. We want something more, than a mere map of the surface of society. We want a deep, intimate, pervading knowledge of the circumstanecs of man's actual condition, and of the influences, whether friendly or adverse, which are acting on his character. Men are divided now into a far greater number of classes, than they have ever before been. We want to know why this is so. All the classes stand much more nearly on the same level, than formerly. Rulers are no longer more, and the ruled are no longer less, than men. The divisions now are not so much of high from low, as of equals from equals. This is a glorious change for the better. We want to understand its nature, cause and extent. But this understanding we have not yet fully attained. We are yet very far from having attained it. And it is this imperfection of our knowledge, rather than any inseparable obscurity belonging to them, that darkens so many questions of deep and vital importance. It is this that makes it so difficult, to point to the cause and effect of a contemporary event, and to decide upon the complexion and tendency of existing circumstances.

Of the questions, relating to the present interests and deeply affecting the present happiness of society, not one, probably, gathers into itself a greater consequence, and certainly, not one ought to excite a livelier concern, than that which we now propose to discuss. This subject is intimately connected with the great topic of human progress. The experiment of machinery has multiplied relations to the condition and prospects of our race. It is a new and almost infinite power, brought to bear on the action of the social system. And, in proportion as it would be consoling and delightful to have reason to believe, that, under the influence of these new impulses, society is advancing and will continue to advance, with swift and constantly accelerating progress, towards the ultimate limits of human improvement; so would it be mortifying, and beyond expression painful, to be driven to acquiesce in the gloomy doctrine, which represents all these new and powerful agents as only working out for man, deeper and deeper wretchedness and degradation.

This question may be more advantageously discussed in our country, than in any other. The experiment of machinery may have a fairer trial here, than elsewhere. The natural course of industry is not obstructed here, in any great degree, by unwise legislation The profits of labor are secured to the laborer. The burthens of taxation are light. The highest motives to exertion operate upon every man in the community. In short, a nearer approach has been made here, than any where else, to a Government that protects all, and injures none; that leaves every one at full liberty to benefit himself, so far as it can be done without injury to others; that takes off every weight and fetter from individual energy, while it restrains all hurtful excesses, and restrains them rather by the fear of public opinion, than the fear of punishment. In such a state of society, every new impulse given to the public mind,—every new agent introduced to further the operations of labor,—exhibits at once its real character and tendency. In such a state of society, the moral action of machinery is not liable to what natural philosophers call the influence of disturbing forces. It operates without restraint, and produces its appropriate effects. It is not complicated with other influences. It has a simple unmodified action of its own, unaffected by the movements around it. Here, therefore, we may ascertain, with comparative facility, what this action is, and what are likely to be its results.

Some of our readers may be surprised, that so much importance and difficulty should be ascribed to this question. They live in the midst of machinery. They see machinery at work on every side, abridging the processes of labor, and making the difficult easy. They are accustomed to regard this subjection of the powers of nature to the will and direction of man, as a splendid triumph of the intellect, and as altogether and unquestionably beneficial in all its tendencies. It is natural, therefore, that they should be astonished when it is made a question. It seems to them quite too plain a matter to admit of argument. But let these persons look abroad. There they will find men, and men too held in high repute for wisdom and honesty, who think and say, that, to those who depend upon their daily labor for their daily bread, or, in other words, to four fifths of almost every nation on the globe, the introduction of labor-saving machinery is a grievous curse. These men will bid them look for a commentary on the influence of machinery, to the condition of the English laborers. They will bid them to ask the half-clothed and half-fed workman, what is his opinion on this subject? They will say to them, 'Inquire of those distracted parents, why they deny food to their famishing offspring? Demand of the whole body of the working classes, what is the cause of this deep, wide-spread distress, which pervades the land like a pestilence, carrying dread to every bosom? Why is Government alarmed? Why is the Church directed to offer up supplications to Heaven, to avert from England the horrible calamities of intestine discord and war? A glance at the condition of the country will answer these questions. There are multitudes of workmen, who either have no employment at all, or labor for wages altogether inadequate to the necessities of existence. This want of employment, and these low wages, are occasioned by the introduction of machinery. The laborers have, of course, become uneasy and discontented. Their irrepressible discontent has at length broken out into open violence. They begin to destroy the machinery. The sure instinct of revenge directs them to the cause of their sufferings. But they do not stop here. They attack the property of their employers, or rather, as they think, their tyrants. Those conflagrations, converting midnight darkness into unnatural day, are their work. There is reason that Government should be alarmed. They are alarmed. They have made strong efforts to arrest the progress of disaffection. The iron arm of power has been stretched out to punish the excesses of these wretched men. Nothing is pardoned to ignorance. Nothing is forgiven to misery. Many have been sentenced to transportation,—many to imprisonment,—many to death. Yet all this avails nothing. Disaffection and disquiet still spread and strengthen. No man is able to foresee what will be the end of these things. 'This,' these honest and intelligent persons will say, 'is in truth a terrible picture of present and impending calamity. But it is only a faint shadow of the real state of things. And if the vengeance of those unfortunate men be not mis-directed,—if, as we believe, machinery be the fruitful mother of all these woes, then, surely, its introduction into such general use cannot be too earnestly deplored.'

It requires no effort of the imagination to suppose this to be the present language of that numerous and highly respectable class of men, who think that the influence of machinery on society is evil and pernicious. There is, however, another class, equally numerous and respectable, who hold the contrary opinion. These persons ascribe the distress, that afflicts the laboring classes of England, and some other portions of Europe, to other and more deep-lying causes, than the introduction of machinery. 'The real springs of all these evils are to be sought for,' they say, 'in vicious political institutions, in unequal laws, and grinding taxation. These are the true fountains, which send forth poisonous and bitter waters. Machinery multiplies the comforts and conveniences of life. Is this an evil? Machinery lightens the burthens of labor. Is relief from the necessity of hard work, a grievance to the laboring classes? No doubt, like every other great power, machinery may be converted into an instrument of great oppression. But it is not such naturally. In itself it has been always, and, under well regulated Governments, it always will be a source of great good,—of good almost unmixed. The evils necessarily incident to its introduction, are slight, partial, and transient. They reach only the surface of society, affect but small portions of the community, and speedily pass away. The benefits arising from the same source, are substantial, universal, permanent. They are seen every where, felt every where, and must abide forever.'

Such is the conflict of opinions on this subject. Where there is so much disagreement, it becomes him, who would share in the discussion, to advance his sentiments with diffidence. We do not dogmatize. We assume the attitude of inquirers, rather than of teachers. We shall be satisfied, though none should be convinced by our labors, if we induce any to examine the subject for themselves,—a subject, which it is important that every man in our community should thoroughly understand. It has not been much discussed, particularly in this country. The general sentiment is decidedly, so far as we have been able to ascertain it, in favor of machinery. A few apostles of the opposite doctrine have arisen here and there; but their converts have not been numerous. Recently, we have observed in some quarters, a disposition to make machinery bear the sins of the tariff; to establish the fact of a partnership between the two, and to make the former responsible for all the faults, real or imaginary, of the latter. We apprehend, that it would not be difficult to demonstrate the absurdity of this notion. We cannot, however, spare more space than is required for the simple statement, that it is groundless. And now, we shall let alone the opinions of others, and proceed to put our readers in possession of our own views on this important subject.

In the earliest ages of society, machinery was unknown.

Man was created in a climate where the earth yielded bountifully at all seasons of the year, her productions for his use. Then, his only labor was to gather from Nature's abundant store, the supply of his present want. Afterwards, he began to cultivate the soil, and then, probably, some simple instrument of culture was invented. At a still later period, his Creator invested him with dominion over the life of living creatures; and, to enable himself to exercise this new authority, he invented, also, rude instruments of hunting and fishing. These are all arts of absolute necessity. Without them, man could not exist, except in the mildest climates, and there only in small numbers. Beyond these arts, a large proportion of the human race have made no great advances. The only additional skill, yet attained by many tribes of the human family,—the skill to make rude clothing and to build rude huts to protect them from the inclemency of the weather,—has been taught them by stern necessity. In their circumstances, each individual of the society must labor for his own subsistence, and all hope of intellectual or moral improvement seems entirely cut off.

From time to time, however, in different parts of the world, there have been communities, which have risen far above this condition. Assyria, Persia, Egypt, Greece and Rome felt, by turns, the genial influence of improvement. And it is worthy of remark, that, wherever, over the whole earth, the light of civilization has once dawned, some rays of that light linger yet,—the utter darkness of absolute barbarism has never returned. It cannot wholly return. The law of man's nature, impressed on him by his God, is onward progress; and let a nation but once rise into the light of civilization, and then, however low adverse circumstances may afterwards thrust them down, they will never sink into utter night, nor will they ever cease to strive to re-ascend to day. It is also worthy of remark, as illustrative of this law, that these nations made different degrees of progress, and that each advanced farther in improvement than the preceding. The light shone faintest, where it first dawned, on Assyria,—its brightest effulgence illuminated Rome. It was a progressive illumination,—faint and hardly perceptible at first,—then gradually receiving greater and greater accessions of splendor. Now, in our day, it has flashed out into a broad, bright, and glorious effulgence, encompassing and illuminating more than half the globe. But of this hereafter.

Our present business is with the cause of all this. Civilization never takes place without the accumulation of the material products of labor. Different causes may produce this accumulation. The hand of violence may gather the spoils of rapine, and manual labor or mechanical contrivance may heap up the store of industry. The two first of these causes, but more particularly the first, procured an abundance of the necessaries and luxuries of life to the states of antiquity; and were, therefore, the principal agents in the work of civilizing those states. Mechanical contrivance exerted a similar, but, at first, almost imperceptible influence, increasing the stores of wealth, and thus helping forward the progress of civilization. The experiment of machinery, however, as a substitute for human labor, employed in producing and increasing the comforts of life, was never tried on a great scale by the nations of antiquity. That was reserved to be the distinction of modern times. The glory of compelling the powers of Nature into the service of man, was destined to grace our own age. And, as the spoils of these bloodless victories have been far greater than ancient conquest ever gained,—as the accumulation of wealth, by the new agents that have been employed in the task, has been far more rapid than was ever known in former times, through the instrumentality of any agent whatever, so civilization has, in these latter days, spread far more widely, and penetrated much more deeply, than ever before; reaching, not one nation only, but many, and bestowing its invaluable benefits, not upon a favored portion merely, but upon the whole of society. We would not say, that machinery has been the only efficient agent of modern civilization. We do not so believe. There have been moral agents at work. They have effected much; but without the aid of machinery, they could not have effected much. What we claim for machinery is, that it is in modern times by far the most efficient physical cause of human improvement; that it does for civilization, what conquest and human labor formerly did, and accomplishes incalculably more than they accomplished. And how different are the characters of these three agents! War, the direst curse of humanity, must necessarily precede conquest; and the structure of civilization, reared by this agent, rises upon the spoil, the desolation, and the anguish of the vanquished. Human labor, when urged to excessive efforts, must necessarily, to a considerable extent, prevent intellectual and moral improvement. But machinery, doing the work, without feeling the wants of man; taking from none, yet giving to all, produces almost unmingled benefit, to an amount and extent, of which we have as yet, probably, but a very faint conception.

There are several objections to this general view of the effects of machinery, which we shall now examine. The first and principal one is, that all labor-saving inventions diminish the demand for human industry, and, consequently, deprive multitudes of laborers of employment. We meet this objection by denying the fact. It is not true, that the demand for human industry is diminished. It is not true, that multitudes of laborers are absolutely deprived of employment. It is true, however, that many laborers are sometimes compelled to change their employment, by the introduction of new and improved machinery into a branch of industry, where a great deal of human labor had been previously required. And it is true, that sometimes, while this change is in progress, a great deal of suffering is experienced. All this we shall attempt to explain.

The earth is the great primary source of the supply of human wants. It is the great laboratory, where the dust we tread upon is converted into life-sustaining nutriment. Whatever we eat, or drink, or wear, comes originally from her bosom. In the earliest stages of society, as has been already said, men consume her productions in their simple state. The springs supply them with water to drink. They eat the fruits of the field, and clothe themselves with leaves and skins. In this savage state, each one supplies his own wants, and it takes all his time to do it. But, after a while, some one more lazy or more ingenious than the rest, discovers some method of lightening his individual labor. Then others imitate him;—and, in time, machines are invented, that seem likely to supersede the necessity of human labor altogether. This would, in fact, be the result, if, in this condition of things, men should consume no more of the products of industry than before; and, of course, a multitude who had been actively employed would be employed no longer. But such is not the fact. The cravings of desire are never satisfied. Extend the supply as you may, the wish for the enjoyments of life will still go beyond it, and will find its only limit in the means of gratification. The only effect, therefore, of increasing the productive energies of labor, by the introduction of machinery, is to distribute it into more numerous departments. A few years ago, those, who roamed through the regions in which we now dwell, exercised, all of them, the same employment. Each one performed his own labor. No one was, in any great degree, dependent on another. How different is the condition of things now! Hardly an individual, of the millions congregated here, produces, himself, the hundredth part of what is required for his own subsistence. The departments of industry are multiplied. The laborers in each are under a tacit obligation to contribute their proportion to the great fund of human subsistence and enjoyment. Each one works for all the rest, and all the rest work for him. In the savage state, all were hunters and fishers; now, some cultivate the ground, some construct machines, some make clothing, some build houses, some make laws, and some preach sermons. Each fills his appropriate place, the amount and the products of human industry are incalculably increased, and the action of the great social system goes on safely and harmoniously.

[The author of "Effects of Machinery" wrote several pages (omitted here) on technological unemployment resulting from the introduction of various labor-saving machines, like spinning jennies and threshing machines. Having conceded that the distress of craftsmen thrown out of work and their families was severe, he then argued that it was short-lived because ultimately the machines lowered prices, increased production, and created new jobs. —Ed.]

It is not at all wonderful, that distress so sore as this should drive men to do what afterwards they are sorry for. Extreme misery impairs the moral sense. The distinctions of right and wrong are apt to be obscured and lost sight of, in the tumult and tempest of passion. Resentment is almost always blind. Its violence generally expends itself on the apparent cause of injury; while it seldom reaches the real cause. It so happened in this case. The laboring classes cried out against machinery, and some statesmen, too, joined in the cry, when the principal source,—by far the most fruitful source,—of calamity, was an imprudent and excessive production, stimulated by high prices. Even this, however, is but a transitory evil. The bright and cheering beams of prosperity are intercepted only by a temporary eclipse. They are not quenched.

They are not extinguished. When production ceases to be profitable, a part of the industry employed in it will be withdrawn. When other employment is found for it, the distress will vanish. And this will take place in a longer or shorter time, according to the circumstances of the nation, and the amount of labor thrown out of employment. In our country, neither the introduction of machinery, nor over-production, can occasion any extensive or permanent evil. The demand for labor is so urgent, that no man need be long out of work. Whenever the current of industry receives a check in one direction, the overflowing waters will immediately find an outlet in another. If machinery bear a part in occasioning distress, it also helps to remove it. If over-production do not irritate the wound, it will soon heal. The man who employs a machine, produces as much as he who employs living workmen. If there be a difference of expense in favor of the machinery, the former will make larger profits than the latter. He will grow rich faster. But he will not put his riches into a strong box. He will surround himself with additional comforts. He will employ a school-master. He will patronize the printer. He will travel and become better acquainted with his race. And thus, while he makes himself a far more useful and valuable member of society than before, he gives employment, in one way and another, to quite as much human industry, as his machine deprived of employment. Thus, from the same cause that produced partial evil, flows also universal good. The amount of productive industry of every sort is, in the end, vastly increased. It has been estimated, that the people of the United States and Great Britain, aided by the improved machinery of the present day, do as much work as could be done by the whole population of the earth, without that aid. And it needs but a glance at the condition of the working classes, (an epithet, which we use for want of one more appropriate to our meaning,) in our country, to convince the candid, that its influence, so far as its ultimate effect on human industry is concerned, is altogether salutary and beneficial.

We have given quite as much attention to the argument against machinery, derived from its effects on labor, as it deserves. It is the strongest and most striking argument that occurs to us on that side of the question, and we wished to state it as fully, fairly, and forcibly as we could. But after all, is it certain that machinery occasions any distress, greater than would have existed, had machinery never been invented? We speak now of that improved machinery, which has, within the last century, so changed the aspect of the civilized world. Before this era, in many countries, the most affluent hardly enjoyed more comforts, than the poorest do now. The poorest classes depended upon servile labor, or an unskillful cultivation of the soil, for a scanty subsistence. They were miserably fed, clothed, and lodged. If they did not feel the wretchedness of their condition so acutely, as men similarly situated would now, it was because none of their neighbors fared much better. They

> Saw no contiguous palace rear its head.
> To shame the meanness of their humble shed;
> No costly lord the sumptuous banquet deal,
> To make them loathe their vegetable meal;
> But poor, and bred in ignorance and toil,
> Each wish contracting bound them to the soil.

But their ignorance was all their bliss. If the thick gloom which involved them, were not a darkness that might be felt, it was because there was no neighboring land of Goshen, where there was light. And to us, their lot seems to be far more worthy of commiseration, because far less susceptible of improvement than that of those, who, at the present day, occasionally experience temporary inconvenience and suffering from want of employment.

The next objection to machinery is, that its tendency is to gather wealth into masses, and widen the distance between the rich and the poor. It is easy to see how this may be the fact in England, where the statute of descents transmits the possessions of the father, almost unimpaired, to the eldest son. The accumulated acquisitions of one generation are handed down to the next, almost unbroken. The eldest son of a rich man must himself be wealthy; and, if he conduct his affairs with prudence, will leave his own eldest son master of a large fortune. The law closes up many of the outlets, by which wealth would otherwise be distributed through the community, and gathers it into the hands of a few. Thus a new order of nobility is created, who have been styled, not inappropriately, the lords of the spinning jenny. It is not machinery, therefore, that widens the distance between the rich and the poor, into an almost impassable gulf, but this law,—a law hardly to be vindicated upon principles of sound policy, under any circumstances, but pernicious and dangerous in the extreme, to a manufacturing and commercial community. In our country, we have no such law, and no such consequences have attended the introduction of machinery. If a rich man, in these States, invest a large fortune in fixed machinery, when he dies, it becomes the property of all his children. Death relaxes the grasp that held the mass of acquisition together, and the law does not put forth its stronger grasp to prevent its natural dissolution. On the contrary, the statute of distribution pulls down the pile of wealth, which the father's industry had accumulated, and divides it among his offspring. It can seldom happen, that there will be enough to make them all rich. The consequence is, that nearly all the individuals of each successive generation, start in the race of life from about the same point; and they are the most successful in that race, who are the most intelligent and the most industrious. It is thus plainly impossible, that, while this statute continues in force, machinery can enrich the few and impoverish the many. Almost all of us have an equal chance to be benefited by its introduction. A machine feels no partialities. It works for one just as vigorously and efficiently as for another. And if any man in this country have no direct interest in machinery, it is simply because he can employ his means more advantageously in some other way. In nearly every instance, where machinery is extensively employed, there is a joint-stock concern. The property is divided into shares, and these shares are held by various individuals. The workmen themselves, who are employed in the manufactory, may, and not unfrequently do, possess an interest in the establishment. It is then little less than absurd to say, that machinery accumulates for the rich alone, while it still farther impoverishes those who are already poor.

A far more serious objection than this remains to be considered. It is alleged, that machinery gathers men together in large masses, confines them in unhealthy apartments, ruins their health, contracts their minds, and depraves their morals; that its wages, like the wages of sin, is death,—moral, intellectual, physical death. This is true in part, and in part it seems to us to be false. It is true, that modern machinery can hardly be used to advantage, especially for manufacturing purposes, without collecting together large numbers of workmen. But it is not true, that these workmen must inevitably be 'crowded in hot task-houses by day, and herded together in damp cellars by night;' that they must toil in unwholesome employments twelve hours a day, and frequently a much longer time; that they must live without decency, and die without hope; that they must sweat night and day, keeping up a perpetual oblation of body and soul, to the demons of gain, 'before furnaces which are never suffered to cool, and breathing in vapors which inevitably produce disease and death.' To all these charges, in behalf of machinery we plead not guilty. They are not true. If they were, well might the genius of humanity be represented as looking on with drooping wings, and a countenance of mingled pity and despair. There would be room for pity. There would be reason to despair. If we admit these allegations to be just, we are driven to the conviction, that the fabric of national greatness, power, and prosperity, however goodly it may seem in its outward show, is but a gorgeous sepulchre, in which are buried the intelligence, the virtue, and the freedom of the mass of the population; that national wealth and national misery go hand in hand, linked together by the strong compulsion of fate, in gloomy yet inseparable companionship. It were better that a nation should remain forever poor and barbarous. Better, far better, that society should make no progress, than that a few should advance and ascend, by treading on the necks of all the rest.

But it is not a necessary, nor a natural consequence of the introduction of machinery, that this state of things should exist. Wherever it does exist, there must be bad laws or a bad Government. We witness no such scenes in our country. The poet, who should search those districts of our country, where machinery is most extensively employed, for images of wretchedness and want, would return disappointed from his quest. The political economist, who should go there for facts to sustain the gloomy theory we have alluded to, would perhaps become a convert to the opposite opinion. There, beautiful villages spring up suddenly, as if the earth had been touched by an enchanter's wand. There, are large and commodious buildings, filled with active machinery, and with intelligent and contented human beings. Around, are their neat and convenient dwellings. There are a few shops, to supply them with a number of little foreign luxuries, which they can well afford to buy; and a tavern, it may be, to furnish, not a resort for the idle and the dissipated, but rest and a temporary home to the weary traveller. There are the schools, in which the children and young persons are instructed how to act well their parts, as free citizens of a free republic. And there, last and best of all, is the church, where, on the sabbath, all, old and young, assemble reverently to worship God. This is no picture drawn from fancy. We have ourselves beheld the real scene, and can attest the verisimilitude of the sketch. And though, in the larger manufacturing towns and cities, a part of these advantages can hardly be enjoyed, yet we may safely appeal to the character and condition of our manufacturing population throughout the whole country, as a standing and unanswerable refutation of the objection, which we have been considering.

We have now done with objections, and will pass to other considerations. We will now say something of the more general effects of machinery; and first, of the vast accession which has been made to the productive energies of labor, and the consequent vast augmentation of the products of industry. The necessaries, the comforts, and the luxuries of life are now produced in unparalleled profusion. The effect of abundant supply is to make articles cheap. Every man can now provide for his wants, and the wants of those dependent on him, in a much easier way, and at a much cheaper rate, than ever before; and the happy consequences of this state of things are visible in the improved condition of all classes of civilized society. But it is not in this point of view, that we chiefly delight to contemplate the effects of machinery. Its influence on the physical condition of man is doubtless very great; but its influence on his intellectual condition is greater. Not only are men in general better fed, better clothed, and better lodged than formerly, but, what seems to us to be a matter of infinitely greater moment, they are far better taught than formerly. Machinery has released some from hand-work, who have applied themselves to head-work. Machinery has supplied them with the means of communicating the results of their industry to the world. Thought is no longer restricted to the narrow circle around the thinker. Machinery has furnished better methods of sending it abroad, than speech. Art has been called in to assist nature. The speaker yields to the writer. The tongue is vanquished by the pen. No power can long preserve, or extensively diffuse spoken words, however eloquent. Write them out and give them to the printer, and if they are worthy of it, they will spread every where and live forever. Formerly, Cicero thundered in the Roman forum, in the midst of the proud monuments of his country's victories, and surrounded on all sides by the altars of his religion, to an audience, that shuddered, and kindled, and quailed, and burned, as he spoke. But when his oration was ended, it was forgotten by the multitude. Some burning thought might be stamped upon the memory. It might pass into a proverb, and be handed down by tradition. But that would not transmit his fame to future times. Had not Cicero written his orations, we should have known little more of him, than that he was a great orator, and that he lived and died in the latter ages of the Roman republic. He did write them, however; but even then, how limited was the circulation, which the copyist alone could give them!

Very different is the case now. A great orator rises in the British parliament. Every word, as it falls from his lips, is caught and written down. Early the next morning, the press gives wings to his thoughts, and sends them abroad, by the aid of the multiplied machinery of conveyance, to traverse regions, and to kindle minds, where Cicero did not even dream that it was possible for man to exist. The epithet, 'winged words,' seems no longer Homeric, but familiar and common-place. In a month's time, they reach New York. In less than a fortnight more, they are descending the Ohio and the Mississippi. In the mean time, they have been passing across the channel, into France, Spain, and Germany, learning new languages as they rush along. They make the circuit of the world. They are heard in India, in Australia, and in the isles of the Pacific ocean. Thus a great thinker and speaker, without the press can do little, against it, nothing; but with its aid, he is like the sun, light radiates from him in all directions, and diffuses itself through space. It may be said of him, without hyperbole, that his words go into all the world, carrying with them a momentous influence for evil or for good. By the side of this tremendous energy, every other power becomes insignificant. It proceeds from mind, and acts upon mind, and it is the chief glory of machinery, that it conveys its impulses to the remotest quarters of the globe. And this power is not conferred only on the great orator and statesman, who stands conspicuously out from the mass of his fellow-men. It is shared, in different degrees, by all who have thoroughly awakened their own immortal energies of mind and spirit. It may emanate from the closet of the poorest student, and be of force to revolutionize an empire. It has been truly, as well as forcibly said, by an illustrious man of our own age and nation, that 'one great and kindling thought, from a retired and obscure man, may live when thrones are fallen, and the memory of those who filled them obliterated, and, like an undying fire, illuminate and quicken all future generations.'

But not only has machinery set free from the necessity of labor, many to teach, but a far greater number to be taught. Let the machines, which now supply the wants of the nations, be destroyed, and it requires no prophetic skill to foresee, that, at the same time, the school-houses will be emptied. Let it be imagined, if any are able to imagine, that every machine, for the furtherance of the operations of labor, is destroyed, and that all memory of the mode of their construction is blotted from the mind, and then let us be told, whether we should not, almost at once, sink back into barbarism. Now, thousands are instructed, where one was formerly. Knowledge is diffused widely through all classes of society, and is yet to be diffused far more widely. An unprecedented demand for useful information is every where made. Through the instrumentality of the press, and the modern engines of swift conveyance, sympathies are established between individuals, and between communities of individuals, who entertain similar sentiments, though residing in opposite hemispheres. It is a remarkable illustration of this, that the friends of freedom and knowledge throughout the globe, seem now to constitute but one great party. Wherever a struggle is made for liberty, wherever a contest is begun with that worst of tyrants, ignorance,—that spot concentrates and fixes the attention of multitudes in every civilized nation. Unnumbered minds watch the progress of the contest, with deep anxiety. Thus a universal public opinion is formed. This opinion has strength in its own nature. It is spiritual, wide-reaching, and mighty. It dethrones kings, it abrogates laws, it changes customs. It is stronger than armies. Barriers and *cordons* cannot shut it out. Fortresses and citadels are no defence against it. It spreads every where, and conquers wherever it spreads.

God grant, that it may continue to spread and to conquer, till every throne of tyranny shall be overturned, and every altar of superstition broken down!

But the most wonderful consequence resulting from the introduction of machinery is, that it has, to all intents and purposes, greatly prolonged the term of human existence. This is not fancy, but fact; not imagination, but reality. Human life should be measured by deeds, rather than by years. He lives long, who accomplishes much; and he lives longer than other men, who accomplishes more than they. And how much more can he accomplish, whose active existence is but beginning now, than was performed, or could have been performed by one, who lived fifty years ago! The multiplied facilities of intercourse, and the cunningly abbreviated methods of doing every thing at the present day, have introduced extraordinary despatch into all the operations of life, and increased a hundred fold the active power of each individual. Whole communities feel the power of these strong exciting influences. Not only are more important and numerous private acts performed by individuals, than have ever before been done in the compass of one life, but public events, more astonishing and of greater consequence, than were wont to happen in former times in the course of centuries, are now crowded into the history of a single generation. And when we look around us, and behold these strong agents of improvement, acting, at the present moment, with greater energy than ever, and producing every day still more wonderful results, we confess, we are filled with astonishment and admiration. We do not claim, as we have already said, for machinery, the sole agency in producing these magnificent effects. We know that they are principally owing to the operation of moral causes. But we say, that it is machinery, which has removed obstructions out of the way of their action, and brought them into contact with the objects on which they are to act; and that, without the aid of machinery, these causes, whatever inherent energy they might possess, could have produced little or no effect on the condition of society.

We have mentioned some of the general results, which machinery has contributed to produce. There is one particular consequence, that should never be forgotten when machinery is spoken of. It was the inventions of two mechanics, that carried England triumphantly through the contest with Napoleon. Arkwright invented a machine for spinning cotton, and Watt perfected the application of steam power to manufacturing purposes. These inventions conquered Bonaparte. They enables Great Britain to manufacture for the world. Wealth flowed into her treasury in copious streams, from every quarter. With this wealth she maintained her own armies, and subsidized those of almost all the nations on the continent. She took the lead in the struggle that ensued, and maintained it, until the battle of Waterloo finally decided the fate of Napoleon and of Europe. These inventions made no great show. They attracted little of popular admiration. No laurels bound the brows of the inventors, though, in our esteem, they were far worthier of the laurel wreath, than the proudest conqueror that ever desolated the earth. Their names have not been blazoned in song. The historian honors them but with a cursory notice. Yet did these men, by their astonishing genius, confer on England power to control the issue of the most momentous and fearful struggle, that ever put in peril the best interests of man.

How does machinery produce these almost miraculous results? How long have these strong influences been acting on society? A few words by way of answer to these questions, shall conclude this article. We have already remarked, that the invention of some machines of a simple construction, is dated far back in remote antiquity; but these were all helps to individual labor, and are never thought of now, when machinery is named. Then, almost every thing was done by hand. Navigation clung timidly to the shore. Labor performed its task tediously and imperfectly. Knowledge was diffused in scanty measures and by tedious processes. Human improvement advanced, if indeed it did advance, imperceptibly. It is but recently, that any great change has taken place. The era of machinery may be said to have commenced within the last fifty years. Man has called upon the unwearied powers of nature to bear his burdens, and they have obeyed the call. Whatever agency expands, contracts, impels, retards, uplifts, or depresses, is set at work. We confine elasticity in our watches, and bid it measure our time. With pulleys and levers, we compel gravitation to undo its own work. We arrest the water as it flows onward to the ocean.

It must do so much spinning, so much weaving, or so much grinding, before it can be allowed to pass on. With the help of pumps and other machinery, we force the very atmosphere we live in, to raise our water from the wells and from the rivers, and to aid in an uncounted and countless variety of other operations. Last, and most wonderful of all, by the application of fire we transform water into that most potent of all agents, steam. Man, as it were, yokes the hostile elements of fire and water, and subjects them to his bidding. It is hardly a metaphor, to call steam the vital principle, the living soul of modern machinery. There is hardly any sort of work, in which this mighty agent may not be employed. It is equal to the vastest operations, and it will perform the most minute. It delves in the mines, it lifts the ore to the surface, and converts it into a thousand forms. It helps to make the engine, which it afterwards inhabits. It brings the cotton to the manufactory, picks it, cards it, spins it, weaves it, stamps it, and then distributes the fabrics for sale. It works on the land and on the water, on the rivers and on the seas. It is found on the Rhine and on the Danube, driving huge fabrics impetuously along through the echoing forests, and by the old castles of chivalrous ages, accustomed to behold far different scenes; and it performs on land the work of many thousand hands. It quickens the activity of commerce on the Indian seas, at the very moment when it is doing the same on the Mississippi and the Ohio. Invention seems to rest from the effort to discover new forces, and to bend all her energies to multiply the applications of this. Friction and gravity alone continue to oppose the dominion of steam over space. Numberless subtle contrivances have been resorted to, to evade the power of these stubborn antagonists of motion. Railways are constructed, stretching over mountains and plains, linking together and making near neighbors of distant territories. Long trains of cars, moving on wheels so peculiarly constructed, that their friction is scarcely perceptible, are placed on them. The horse is unharnessed. He is too slow and too weak to perform the required service. At command, the whole moves, hurrying on, under the strong impulse of an invisible power, with a velocity that defies description. The lover need no longer pray for wings to bear him through the air. A railway car will bear him swifter than the swiftest wing. The exclamation of the poet no longer startles us. His description of the physical achievements of man's 'genius, spirit, power,' are no longer extravagant. It falls far short of the reality.

Look down on Earth!—What seest thou?—Wondrous things!
Terrestrial wonders that eclipse the skies!
What lengths of labored lands! What loaded seas!
Loaded by man for pleasure, wealth or war!
Seas, winds and planets, into service brought,
His art acknowledge, and subserve his ends.
Nor can the eternal rocks his will withstand;
What levelled mountains! and what lifted vales!
High through mid air, *here*, streams are taught to flow;
Whole rivers, *there*, laid by in basins, sleep.
Here, plains turn oceans; *there*, vast oceans join
Through kingdoms, channelled deep from shore to shore.
Earth's disembowelled! Measured are the skies!
Stars are detected in their deep recess!
Creation widens! Vanquished Nature yields!
Her secrets are extorted! *Art* prevails!
What monument of genius, spirit, power!

This was a just description when it was written; and it describes splendid triumphs of the intellect over matter. Let our readers add to it all the wonders which have been achieved by steam, and they will have a tolerably accurate idea of what the mechanical powers have done, and are doing for man.

[A passage from Cicero's *De Natura Deorum* describing man's dominion over nature in his day has been removed. —Ed.]

INTRODUCTORY NOTE

Olmsted's primary objective in "On the Democratic Tendencies of Science" was special pleading for the American colleges and universities, but the nature of his plea illuminates an American attitude toward science and technology. Intending to counter the argument that the colleges and universities were aristocratic institutions designed chiefly for the benefit of the rich, Olmsted insisted that their doors were open to all classes. Young men so educated, Olmsted continued, could contribute to invention through their scientific understanding. And it was "the inventions of science" that tended "to elevate the masses, and to produce social equality."

The most revealing aspect of the Olmsted essay for the student of American attitudes toward technology is his insistence that the major inventions of his era—telegraph, steamboat, locomotive, and labor-saving devices—were producing social equality and were "rendering the conveniences and elegancies of life accessible to the many instead of the few." His exposition is a good example of the common argument of the first half of the century that democracy and useful science, or technology, are mutually reinforcing.

Olmsted's belief that the great inventions of his period stemmed from the work of the philosophers and men of science, however, runs counter to the view more generally held today that the majority of the early inventors were practical mechanics and self-trained engineers unschooled in the higher sciences. It would be difficult, if not impossible, to support Olmsted's assertion that Morse, Fulton, and Whitney were philosophers and men of science and that the railway engineers of England were true scientists.

Denison Olmsted (1791–1859) at an early age became interested in teaching and hoped, after receiving his degree at Yale College, to establish a normal school. However, he accepted a professorship of chemistry at the University of North Carolina and thereafter taught science and mathematics. In 1825 he was called to Yale. He conducted scientific investigations, spoke and wrote on education, invented practical devices, and wrote widely used science textbooks. The essay reprinted here reflects his concern for pedagogy, science, and technology.

On the Democratic Tendencies of Science

DENISON OLMSTED

Read before the American Association for the Advancement of Education.

It has been but too common in our country to raise an outcry against Colleges and Universities, as being aristocratic institutions, designed chiefly for the benefit of the rich. The same charges have sometimes been brought against Science itself, as tending to produce and perpetuate invidious distinctions among men, giving to the few undue ascendency over the many. Under this idea, legislatures have thought it necessary to confine all appropriations of money for the benefit of education to the common schools, on the ground that the higher institutions of learning are not for the benefit of the people at large, but only for the wealthy and privileged classes of society. Demagogues, also, have found a fruitful theme in declaiming against colleges and universities, as institutions intended chiefly for rich men's sons, and therefore they have claimed to be the friends of the people by espousing, exclusively, the cause of the common schools, and preventing, as far as lay in their power, any of the State funds being applied to sustain the higher institutions of learning.

My object, in the present essay, is to prove that science, in its very nature, tends to promote political equality; to elevate the masses; to break down the spirit of aristocracy; and to abolish all those artificial distinctions in society which depend on differences of dress, equipage, style of living, and manners; to raise the industrial classes to a level with the professional; and to bring the country, in social rank and respectability, to a level with the city. In support of this doctrine, I shall endeavor to show that such is the whole drift and tendency of science, first, in its inventions, and secondly, in its institutions.

Barnard's Journal of Education, 1 (1855–1856), 164–173.

1. THE INVENTIONS OF SCIENCE TEND TO ELEVATE THE MASSES, AND TO PRODUCE SOCIAL EQUALITY

Such, I aver, has been the actual effect of the changes which the inventions of science have brought about in our own country within the last fifty years—a period distinctly within my own recollection. These changes have been chiefly effected in the following ways: first, by improvements in the arts of *locomotion;* secondly, by the general *diffusion of intelligence,* especially through the medium of newspapers; thirdly, by an extraordinary multiplication and cheapening of the *conveniences* and *elegancies of life.* Let us review each of these particulars separately, and then consider how far they are due to the labors of science.

We will first look at the effects of *steamboats* and *railroads* in producing social equality. My remarks will be understood to refer chiefly to what I have witnessed in Connecticut—a district to which my field of observation has been for the most part limited. Before the introduction of steamboats and railroads, there was a great distinction maintained between the professional and the industrial classes, and between men of wealth and what were called the common people, in their respective modes of traveling. Men of wealth kept their carriages with their drivers. In these their families took their rides about the town, and in these they made their journeys abroad. Meanwhile the laboring classes, such as farmers or mechanics, jogged along in plain, unornamented, rattling wagons, or rode on horseback. The gentlemen in coaches were looked up to as a superior class of people, with whom those in wagons or on horseback could not presume to claim any acquaintance, or to have any except the most formal intercourse; and those in coaches claimed the privileges of caste, and expected a deference from the other party, corresponding to the difference in their equipages, or if they spoke to them at all, considered it an act of great condescension. Merchants, when on errands of business, generally rode in the public stages; but this mode of traveling was too expensive for the farmers and mechanics, and was little used by them. Indeed, people of these classes seldom had occasion to go so far from home as to require the accommodation of the public stages; and since they had little intercourse with the educated, and professional, and wealthy classes, in the daily relations of life at home, and still less abroad, the two classes of society recognized as the upper and lower classes, had as little intercourse with each other as though they had been separated by the odious distinctions of caste. Merchants, by frequent visits to the cities, acquired somewhat of the manners of the city, and adopted a style of building, furniture, and dress, which distinguished them from the farmers and mechanics, as much as the professional were distinguished from the industrial classes. The term "countrified" was an epithet of reproach liberally applied by the inhabitants of the cities, even of the smaller cities, to the country people, who, again, conscious of their ignorance of the forms of genteel society, and of the rusticity of their clothing, felt abashed when they came into the presence or entered the houses of the city-bred people. It was my fortune (I do not say *mis*-fortune) to be country-bred, and I well remember my visit, when a boy, to the neighboring city, mounted on a nag whose mane and tail were not trimmed after the city fashion, a pack of boys following me, throwing missiles, and hallooing "Country!"

If we now enter the saloon of a steamboat where the passengers, male and female, are assembled in great numbers, we shall probably be in the midst of people of many different situations in life, varying widely in education and fortune, some city-bred and some country-bred, representing many different professions—the learned and the industrial—mechanics, farmers, lawyers, merchants, clergymen, physicians, judges, statesmen, teachers, with the wives and daughters of each and every class. Yet the people who compose this promiscuous assemblage will differ so little in general appearance and manners, that we shall feel puzzled to assign the peculiar vocation of any, much less to determine which belong to the higher and which to the lower class in society. In fact, this anti-republican distinction is nearly obliterated in our State, and the separation is not now into the upper and lower classes, but into the virtuous and the vicious, the industrious and the indolent, the temperate and the intemperate.

If we enter a railroad car, we may again meet with people of many different vocations, but we recognize no appearance of caste. All mingle together on terms of perfect reciprocity. The intimate contact into which people of different professions are brought in the rail cars is working most salutary changes in the sentiments of different portions of society toward each other. The scholar takes his seat, unconsciously, by the farmer or the mechanic; they enter into free conversation, first upon topics of common interest, as the weather or the news of the day, but afterward on subjects appropriate to each. The scholar learns of the farmer and the farmer of the scholar, and each makes a grand discovery—the scholar, that the farmer is not half so ignorant as he had supposed, and the farmer, that the scholar is not half so proud as he thought he was. Mutual respect is the consequence, and the desire of a more extended intercourse between people of different professions is increasing, to the mutual benefit and respect of both parties. By the facility with which visits are now paid to the large cities, the people of the country resort to the cities much more than formerly. By this more enlarged intercourse with refined society, the characteristics of provincialism are fast wearing away. The countryman is no longer detected by the coarse texture or rustic fashion of his coat, or the uncouthness of his manners, or the peculiarities of his dialect and pronunciation. The refinements of taste, also, are rapidly spreading over the interior. Handsome houses, genteel furniture, and refined habits of living, have made wonderful progress in the interior of our State within a few years. There is scarcely a village in Connecticut where we may not find families living as genteelly as the better class of families lived in the city of New York fifty years ago.

Not only has there been great progress all over the country within the period of steamboats and railroads, in a taste for the embellishments of art and the refinements of civilized life, but the steamboats and railroads have themselves furnished the means of gratifying that taste. They have enriched the country by greatly enhancing the value of its production, both mechanical and agricultural. How have they opened to this generation the exhaustless riches of the Mississippi Valley, and filled all New England with thriving manufactories!

We will next contemplate the changes which have occurred within the last fifty years in the general and rapid diffusion of intelligence among the industrial classes. Within my recollection the progress of a piece of foreign news, from the metropolis to the interior of Connecticut, was something like the following: The New York papers containing it traveled slowly in the stages, stopping over-night, until in the space of two or three days it reached Hartford. Then in the course of a week it was republished in one of the weekly papers, of which there were two, but none were issued oftener than once a week. A post-rider, on horseback, distributed this paper among the country people, several farmers in one neighborhood frequently making a single paper a joint stock concern. From two to four weeks generally elapsed before an article of news reached the heart of New England, after it was first known in New York or Boston; and a very large proportion of the inhabitants took no newspaper, and hardly received the tidings in any way, except by an indefinite rumor. Steamboats first gave an increased speed and range to newspapers, and at a later day railroads have so augmented both, that there is scarcely a village in New England where the New York morning papers are not read before night on the same day. Moreover, with the means of indulgence, the appetite for news has been wonderfully excited, so that a daily newspaper from New York or Boston, or issued in the town, has become to almost every man in New England one of the necessaries of life. The consequence is, that the country people are no longer looked down upon by the people of the large cities as "behind the times;" as knowing nothing of what is going on in the world, since a few hours only intervene between the merchant on change and the farmer at the plow, in the remotest parts of New England. The effect thus begun by steamboats, and continued by railroads, in elevating the country to an equality, in social condition, with the city, the *telegraph* has completed. In no important piece of intelligence is the country—east, west, north, or south—more than a few hours, seldom more than a few minutes, behind the metropolis. In no respect is the equality of the country and the city, produced by the inventions of science, more conspicuous than in this. In places where but thirty years ago the untamed savage or the wild beast roamed, in the remote districts of the West, the arrival of an Atlantic steamer at New York, or the results of the morning stock-board are matters of familiar conversation within two or three hours after they are first known at the Merchants' Exchange.

We will take but one example more to illustrate the great change in the social condition of all classes of the American people, which the last half century has produced, and that respects the effects of science in rending the conveniences and elegancies of life accessible to the many instead of the few. It must be obvious to every observer, that we of the present generation feed on better fare than our fathers did, wear vastly finer and better clothing, live in far better houses, and enjoy infinitely more of the comforts and even the luxuries of life, to say nothing of the embellishments of taste (which formerly were exclusively within the reach of the rich and great), and with all this we do not labor half so hard as our fathers labored.

The facts which have been adduced are sufficient to show that *something* has, within the last half century, greatly elevated the privileges and enjoyments of the masses of our countrymen, and produced a far greater equality in the social condition of the laboring, in comparison with the wealthy classes, and vastly augmented the intelligence and respectability of the country, in comparison with the city. Now, the only question we have to examine is, has *science* done it? I do not say that science would have done it, to the same extent, except in a free country, enjoying all the blessings of a free government; but in *our* country I do say that these happy changes have been the true and legitimate results of science.

We have seen that the changes described have been the immediate results of steamboats, and railways, and the magnetic telegraph, and improvements in manufactures, by means of labor-saving machines, and the introduction of various chemical arts. But how came society in possession of steamboats, and railways, and locomotives, and telegraphs? Who have chiefly been the inventors of the labor-saving processes which have secured such cheapness to the comforts and elegancies of life, as to place them within the reach of every man of moderate fortune, whereas before, those who wore fine linen were only the rich and the noble? Who invented the steam-engine itself? Watt, a philosopher, a man of science. Who applied it to steamboats? Robert Fulton, a man thoroughly versed in the science of mechanics. Who applied it to railroads? The scientific engineers of England. Who invented the electric telegraph, by which the country is raised to an equality with the city? It was Morse, a son of Yale. Who invented the cotton-gin, by means of which, not only have the cotton planters been enriched, but every one who wears a cotton garment derives benefit from the invention, in the cheapness of the article? It was Eli Whitney, another son of Yale. Who have substituted the modern art of bleaching—the work of a day—for the slow, tedious, and expensive methods formerly practiced, and have thus cheapened clothing, and helped to reduce the price of fine fabrics, so as to bring them within the reach of everybody, and have contributed greatly to reduce the price of writing and printing paper, and thus to promote the general diffusion of knowledge by books and newspapers? This immense improvement in the art of bleaching was a present which Chemistry made to the arts. Mineralogy and Geology also have contributed their share, by laying open new beds of coal, for feeding the fires by which the steamboat and the locomotives are impelled; and Chemistry and Natural Philosophy unite their powers in investigating the laws of heat, and in contriving apparatus to render its applications most effective and economical.

Some will acknowledge that a few men of science, of a practical turn of mind, have contributed to the elevation of the masses by their useful inventions, while they can not see how men who are pursuing science in the abstract, as it is taught in our colleges and universities, are doing any thing for the general good. But if we go back one step beyond the inventors themselves, we come to the original investigators of the *principles* from which their inventions sprang. In the steam-engine we must go back of Mr. Watt to Dr. Black, the chemist, who investigated the laws of steam, without a knowledge of which it could never have been successfully employed as a mechanical force. In the telegraph, we must go back of Mr. Morse to Franklin and others, who discovered the properties and laws of electricity. If we admit that Hadley, a philosopher, presented the sailor with his quadrant, we must not forget that back of Hadley was another philosopher of the closet, who developed the optical principles upon which the quadrant depends. If it is granted that he who calculated the nautical tables, by which the mariner finds his place on the ocean, is a practical man, it must be granted also that the mathematician is a practical man, who furnished the calculator with his rules, and still more the astronomer, who determined the motions of the heavenly bodies, upon which the tables are founded, and, most of all, Newton and Laplace, who discovered and developed the great principle of gravitation, that enabled the astronomer to fix so accurately the places of the heavenly bodies. Thus science, in its very nature and in all its forms, whether cultivated by the recluse philosopher in his laboratory, or applied immediately to the wants of society, in the form of useful inventions, tends to equalize the gifts of Heaven, and to produce social equality among men.

2. THE INSTITUTIONS, NO LESS THAN THE INVENTIONS OF SCIENCE, TEND TO ELEVATE THE MASSES AND TO PRODUCE SOCIAL EQUALITY

It is no doubt true that some of the universities of Europe, under absolute governments, or amid powerful aristocracies, confer peculiar privileges on the sons of the nobility; but in the United States we neither know nor acknowledge any such order, and nowhere in our country are the accidents of birth and fortune less thought of than in our colleges. In what I say on this subject, I shall, indeed, have more particular reference to Yale College, where I have had full opportunity for observation for a period of more than forty years; but, no doubt, most of my remarks will be applicable alike to all our higher seminaries of learning.

In the first place, nearly all our older American colleges are charitable institutions, founded and sustained by the contributions of the pious and benevolent; and if among them there have been some men of wealth who, either during their lives or at their death, have given largely to such institutions out of their treasures, yet they have always, it is believed, been of the number of those who have least desired to promote colleges for the exclusive benefit of rich men's sons. The cause of useful knowledge, the general elevation of society, the interests of the Redeemer's kingdom—these are the motives which have generally, if not always, influenced those who have endowed colleges.

In the second place, the *terms* on which our colleges offer an education are fixed at the lowest possible rate, in order that men of small means may have the opportunity of educating their sons. At Yale College, the rate of tuition is fixed at a price much lower than is paid in academies and private seminaries of learning; and from this low rate there are numerous instances, in cases of pressing indigence, where a part or the whole price of tuition is abated. Moreover, there are funds held in reserve for the express purpose of enabling poor men's sons, of fair promise, to secure the blessings of a liberal education.

It is for the same great object, namely, that the college may have the power of aiding by its funds indigent young men, that the salaries of its officers are fixed at a rate adequate only to a bare support, and often, indeed, below what is required for the support of a family on a moderate scale of respectability. If there is any one point where, at present, the colleges of New England are more especially emulous of each other, than on any other point, it is in affording the greatest encouragement to indigent young men. We have opened to such candidates so many opportunities for helping themselves, and released them from paying the college bills to such an extent, that every year examples* are afforded to students who have passed through college, and fully shared in its advantages, without any resources beyond their own earnings.

In the third place, it is not rich men's sons, as a class, that enjoy at our colleges the greatest measure of respectability, but it is the sons of farmers, mechanics, clergymen, and other men of moderate means; and, in fact, frequently among the most respected are those who, in order to pay their expenses, do every sort of work which they can obtain, such as ringing the bell, sawing wood, and taking care of the public rooms. Nor, in the distribution of college honors and distinctions, is the question ever raised whether the candidate is country-bred or city-bred; whether he is the son of a rich man or a poor man; whether his father is a high officer of state or in a menial condition. And since the sons of the industrial classes are usually brought up to greater habits of industry, and with a higher appreciation of the value of time, the students of those classes do in fact share more largely in the college honors and distinctions, and enjoy a higher degree of consideration than the sons of the rich and great. I venture to repeat, that nowhere on earth are men estimated more exactly according to their true merit, independently of all considerations of family or fortune, than at Yale College.

But, in fact, our colleges are not, as is supposed by some, made up of rich men's sons. Without pretending to be very accurate, I would, for a general idea, distribute the students of

*These, however, are to be considered as remarkable examples of talents united with enterprise; to be very destitute is, in most cases, a great embarrassment and affliction to the student, and sometimes seriously impairs his scholarship.

Yale College into the four following groups: The first quarter may consist of the sons of the decidedly rich, although this I believe to be much above their true proportion. The second quarter may be allotted to the better half of the middling class, who, although not accounted rich, are able without inconvenience to pay the expenses of their sons' education. The third quarter may be assigned to the lower half of the middling class, sons of substantial farmers and mechanics, who, nevertheless, find themselves somewhat straitened to meet the expenses of their sons at college. The fourth class may be composed of such as are decidedly indigent, who work their way through college by a severe economy united with various self-denying expedients for defraying their expenses, and, in many cases, come to the end of the race with a considerable debt upon their shoulders. Foreign universities may abound with sons of the nobility, but to represent our American colleges as institutions devoted to the rich is false in fact.

In the fourth place, if we now follow the men educated at our colleges into life, and view them on the great field of action, it will not appear that the sons of the rich are particularly prominent above the sons of the poor. I apprehend it will be found that, as a class, they make a less figure than either of the other quarters into which we supposed the whole to be distributed. If, by the aid of the triennial catalogues of our colleges, we endeavor to ascertain who among successive college classes have become most eminent, I think they will prove to be those who have come from the industrial professions, or from families who are alike removed from great wealth and excessive indigence, although there are occasionally striking exceptions in both extremes. Or if, instead of endeavoring to form this comparison by so imperfect a guide as the triennial catalogues, we look abroad upon the face of society itself, and see who are actually occupying the posts of usefulness and have attained the highest stations of eminence in church and state, we shall be convinced of two facts: First, that the men who are at this moment exercising the greatest influence in society, in the cabinet of the United States, in the halls of Congress, on the bench of justice, in the State governments, divines, physicians, lawyers, instructors of youth, are, in great proportion, such as have been trained at the higher seminaries of learning; and, secondly, that these have, in a majority of instances, ascended from the classes of society which lie below the wealthy class. What has made them what they are? What has taken them from the obscurity in which they were born, and given them such ascendency in the Republic? What but these very colleges and universities, which are denounced by demagogues and neglected by legislatures, as institutions which are designed chiefly for the benefit of the rich, while the common schools only are deemed worthy of legislative patronage, as institutions which confer their benefits on the people at large —on the *many* in contradistinction to the *few*.

Nothing, again, is more unjust to the higher seminaries of learning, than to represent them as the enemies of popular instruction. Their sons, whenever they have a voice in legislation, are almost always the most liberal promoters of popular education, and labor most assiduously in behalf of the common schools; and the colleges themselves have the highest interest in elevating the standard of popular education, for it is from the more intelligent portions of the community that they derive both their funds and their pupils.

I will only add, that I look upon all the institutions of learning—the common-school, the academy, the normal school, and the university—as acting and re-acting on each other like the grand powers of nature, and all as deserving of the highest possible aid from every enlightened government.

PART III

The Realization of Technological Power

INTRODUCTORY NOTE

In the late nineteenth century, the subject matter of *Scientific American* included applied science, engineering, and invention. It was, in fact, the leading journal reporting technological developments in a manner precise enough for the professional and sufficiently comprehensible for the interested general reader. Its handsome and informative illustrations were one manifestation of the well-developed state of technology about which the journal reported.

The article reprinted here was a *Scientific American* prizewinner. To celebrate its semicentennial, the journal asked for essays on the topic "The Progress of Invention During the Past Fifty Years." The prize-winning entry quite appropriately stressed the steep rate of increase in the issuance of American patents during the half-century. Not only had the number increased dramatically, but it had forged ahead of comparable statistics for the rest of the world. This must have caused many Americans to see themselves as the world's leading technological power—a heady notion for a relatively new nation.

"Beta" interpreted this outpouring of patentable inventions as the manifestation of a golden era of human history. He was not restrained in his superlatives as he associated patents with the "sacred" quality of creation. Nor did he hesitate before attributing this characteristic to "a relatively small number of the Caucasian race under the benign influences of a Christian civilization." Unique American institutions and laws also explained American inventive genius. For him, the preeminence of Caucasians was not ephemeral, nor did he anticipate a drying up of the creative wellspring. He assumed that progress and the facts of evolution taught that "human ingenuity knows no limit." Even though the enthusiasm of Beta seems exaggerated decades later, his essay reflects an attitude common in his time.

Edward W. Byrn (1849–1921) is the author of *The Progress of Invention in the Nineteenth Century* (New York, 1900).

The Progress of Invention During the Past Fifty Years

"BETA" (EDWARD W. BYRN)

If the life of man be threescore years and ten, fifty years will about mark the span of ripe manhood's busy labor, and the sage of to-day, turning back the pages of memory, may, as the times pass in review, enjoy the rare privilege of personal observation of, direct contact with, and positive knowledge concerning the events of this prolific period. To him what a vista it must present; what a convergence of the perspective; for the past fifty years represents an epoch of invention and progress uniqe in the history of the world. It is something more than a merely normal growth or natural development. It has been a gigantic tidal wave of human ingenuity and resource, so stupendous in its magnitude, so complex in its diversity, so profound in its thought, so fruitful in its wealth, so beneficent in its results, that the mind is strained and embarrassed in its effort to expand to a full appreciation of it. Indeed, the period seems a grand climax of discovery, rather than an increment of growth. It has been a splendid, brilliant campaign of brains and energy, rising to the highest achievement amid the most fertile resources, and conducted by the strongest and best equipment of modern thought and modern strength.

The great works of the ancients are in the main mere monuments of the patient manual labor of myriads of workers, and can only rank with the buildings of the diatom and coral insect. Not so with modern achievement. This last half century has been peculiarly an age of ideas and conservation of energy, materialized in practical embodiment as labor-saving inventions, often the product of a single mind, and partaking of the sacred quality of creation.

The old word of creation is, that God breathed into the clay the breath of life. In the new world of invention mind has breathed into matter, and a new and expanding creation un-

Scientific American, 75 (July 25, 1896), 82–83.

folds itself. The speculative philosophy of the past is but a too empty consolation for short-lived, busy man, and, seeing with the eye of science the possibilities of matter, he has touched it with the divine breath of thought and made a new world.

It is so easy to lose sight of the wonderful, when once familiar with it, that we usually fail to give the full measure of positive appreciation to the great things of this great age. They burst upon our vision at first like flashing meteors; we marvel at them for a little while, and then we accept them as facts, which soon become so commonplace and so fused into the common life as to be only noticed by their omission.

Perhaps, then, it will serve a better purpose to contrast the present conditions with those existing fifty years ago. Reverse the engine of progress, and let us run fifty years into the past, and practically we have taken from us the telegraph, the sewing machine, the bicycle, the reaper and vulcanized rubber goods. We see no telephone, no cable nor electric railways, no electric light, no photo-engraving, no photo lithographing nor snapshot camera, no gas engine, no web perfecting printing press, no practical woodworking machinery nor great furniture stores, no passenger elevator, no asphalt pavement, no steam fire engine, no triple expansion steam engine, no Giffard injector, no celluloid, no barbed wire fence, no time lock for safes, no self-binding harvester, no oil nor gas wells, no ice machines nor cold storage. We lose the phonograph and graphophone, air engines, stem winding watches, cash registers and cash carriers, the great suspension bridges, iron frame buildings, monitors and heavy ironclads, revolvers, torpedoes, magazine guns and Gatling guns, linotype machines, all practical typewriters, all pasteurizing, knowledge of microbes or disease germs, and sanitary plumbing, water gas, soda water fountains, air brakes, coal tar dyes and medicines, nitroglycerine, dynamite and guncotton, dynamo electric machines, aluminum ware, electric locomotives, Bessemer steel, with its wonderful developments, ocean cables, etc. The negative conditions of that period extend into such an appalling void that we stop short, shrinking from the thought of what it would mean to modern civilization to eliminate from its life these potent factors of its existence.

As the issue of patents in this country is based upon novelty, it will aid us in the effort to appreciate this great move-

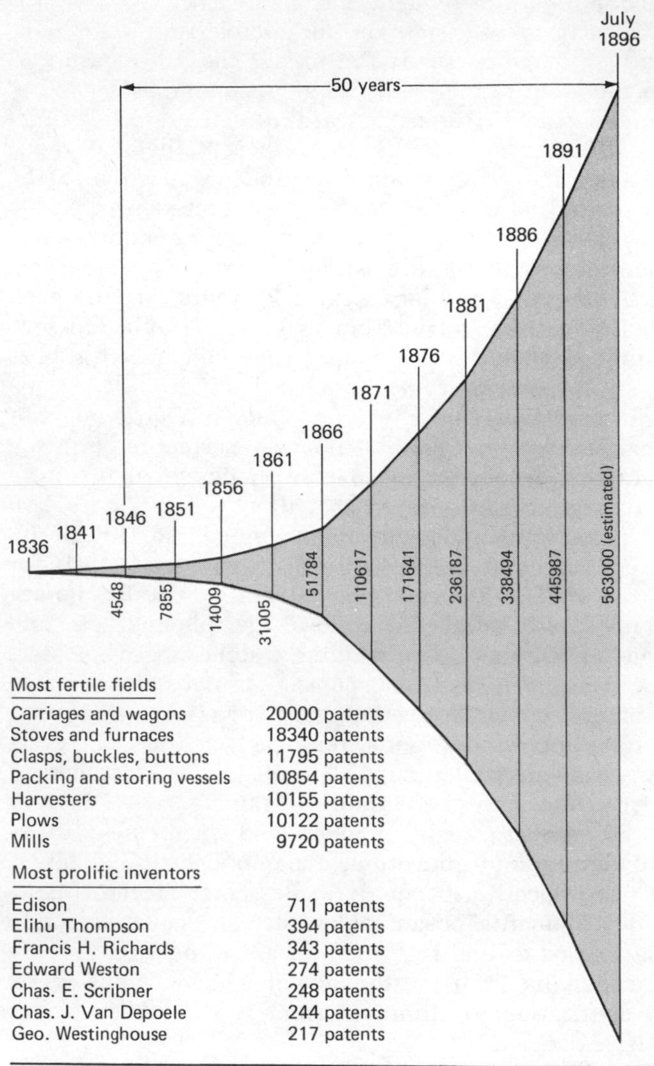

Diagram 1. Diagram showing ratio of increase of U.S. patents for each 5 years.

ment to note the increase of United States patents in the past fifty years. Beginning in 1846, and dividing the time into periods of five years, the increase is shown most graphically in the scaled Diagram 1.

If the growth of United States patents and the progress of the last half century can be taken as fairly correlated, what an insignificant thing is the little attenuated triangle back of 1846 compared with the swelling curves of the later period! It is probably safe to say that fully nine-tenths of all the material riches and physical comforts of to-day have grown into existence in the past fifty years.

It is interesting to observe how closely the grant of patents and the prosperity of the country are related. Referring to scaled Diagram 2, the zigzag line marks the increase or decrease in the patents issued from year to year. We note the depression of the civil war, followed by the rapid reaction and growth of reconstruction. Again, the depression caused by the financial panic of 1873, and again in 1876, the unsettled and dangerous condition of politics incident to the contested presidential election. This was followed by another wave of pros-

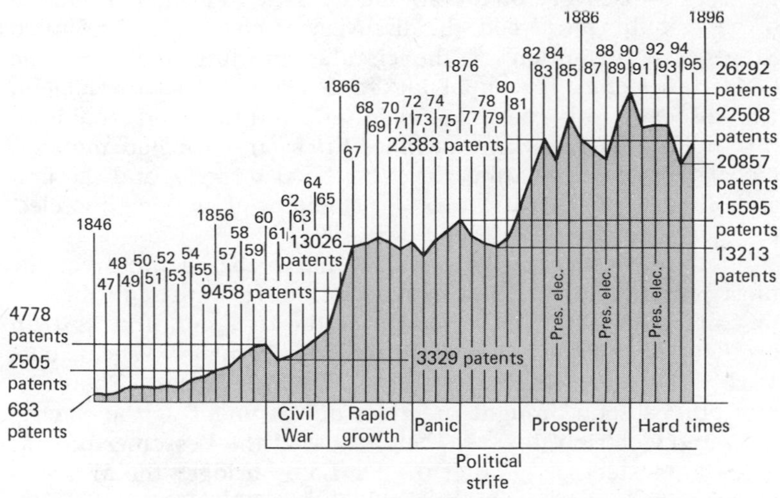

Diagram 2. Diagram showing increase or decrease of U.S. patents from year to year.

perity, indented with depressions in the presidential election years, while the stringency of the times from 1890 to 1894 shows a marked influence in the corresponding depression in the line, all of which indicates a most sympathetic relation.

Passing now to the chronological development of the period, Morse had just harnessed the most elusive steed of all Nature's forces, and put it in the permanent service of man; when nitro-glycerine, discovered by Sobrero, in 1846, for the first time lent its terrible emphasis, and seemed to bring an awakening of the dormant genius of man.

Within the first decade (1846–1856) came the sewing machine, Bain's chemical telegraph, the Suez Canal, the House printing telegraph, the McCormick reaper, the discovery of the planet Neptune, the Corliss engine, the collodion and dry plate processes in photography, the Ruhmkorff coil, the Bass time lock for safes, the electric fire alarm of Channing & Farmer, Gintle's duplex telegraph, the sleeping car of Woodruff, Wilson's four-motioned feed for the sewing machine, Ericsson's hot air engine, the Niagara suspension bridge, and the building of the Great Eastern.

The next decade (1856–1866) brought with it the Atlantic cable, the discovery of the aniline dyes by Perkin, the making of paper pulp from wood, the discovery of coal oil in the United States, the invention of the circular knitting machine, the Giffard injector, for supplying feed water to steam boilers; the discovery of caesium, rubidium, indium and thallium; the McKay shoe sewing machine, Ericsson's ironclad monitor, Nobel's explosive gelatine, the Whitehead torpedo, and the first embodiment of the fundamental principles of the dynamo electric generator by Hjorth, of Denmark.

The next decade (1866–1876) marks the beginning of the most remarkable period of activity and development in the history of the world. The perfection of the dynamo, and its twin brother the electric motor, by Wilde, Siemens, Wheatstone, Varley, Farmer, Gramme, Brush, Weston, Edison, Thomson, and others, soon brought the great development of the electric light and electric railways. Then appeared the Bessemer process of making steel; dynamite; the St. Louis bridge; the Westinghouse air brake; and the middlings purifying and roller processes in milling. That great chemist and probably greatest public benefactor, Louis Pasteur, added his work to this period;

the Gatling gun appeared; great developments were made in ice machines and cold storage equipments; machines for making barbed wire fences; compressed air rock drills and the Mont Cenis tunnel; pressed glassware; Stearns duplex telegraph, and Edison's quadruplex; the cable car system of Hallidie, and the Janney car coupler; the self-binding reaper and harvester; the tempering of steel wire and springs by electricity; the Lowe process for making water gas; cash carriers for stores; and machines for making tin cans.

With the next decade (1876–1886) there arose a star of the first magnitude in the constellation of inventions. The railway and telegraph had already made all people near neighbors, but it remained for the Bell telephone to establish the close kinship of one great talkative family, in constant intercourse, the tiny wire, sentient and responsive to the familiar voice, transmitting the message with tone and accent unchanged by the thousands of miles of distance between. Then come in order the hydraulic dredges, and Mississippi jetties of Eads; the Jablochkoff electric candle; photography by electric light; the cigarette machine; the Otto gas engine; the great improvement and development of the typewriter; the casting of chilled car wheels; the Birkenhead and Rabbeth spinning spindles; and enameled sheet iron ware for the kitchen. Next the phonograph of Edison appears, literally speaking for itself, and reproducing human speech and all sounds with startling fidelity. Who can tell what stores of interesting and instructive knowledge would be in our possession if the phonograph had appeared in the ages of the past, and its records had been preserved.

The voices of our dead ancestors, of Demosthenes and Cicero, and even of Christ himself speaking as he spake unto the multitude, would be an enduring reality and a precious legacy. In this decade we also find the first electric railway operated in Berlin; the development of the storage battery; welding metals by electricity; passenger elevators; the construction of the Brooklyn bridge; the synthetic production of many useful medicines, dyes, and antiseptics, from the coal-tar products; and the Cowles process for manufacturing aluminum.

In the last decade (1886–1896) inventions in such great numbers and yet of such importance have appeared that selection seems impossible without doing injustice to the others. The graphophone; the Pullman and Wagner railway cars and vestibuled trains; the Harvey process of annealing armor plates; artificial silk from pyroxyline; automobile or horseless carriages; the Zalinski dynamite gun; the Mergenthaler linotype machine, moulding and setting its own type, a whole line at a time, and doing the work of four compositors; the Welsbach gas burner; the Krag-Jorgensen rifle; Prof. Langley's aerodrome; the manufacture of acetylene gas from calcium carbide; the discovery of argon; the application of the cathode rays in photography by Roentgen; Edison's fluoroscope for seeing with the cathode rays; Tesla's discoveries in electricity, and the kinetoscope, are some of the modern inventions which still interest and engage the attention of the world, while the great development in photography, and of the web perfecting printing press, the typewriter, the modern bicycle, and the cash register is beyond enumeration or adequate comment.

Looking at this campaign of progress from an anthropological and geographical standpoint, it is interesting to note who are its agents and what its scene of action. It will be found that almost entirely the field lies in a little belt of the civilized world between the 30th and 50th parallels of latitude of the western hemisphere and between the 40th and 60th parallels of the western part of the eastern hemisphere, and the work of a relatively small number of the Caucasian race under the benign influences of a Christian civilization.

Remembering, furthermore, that most of this great development is of American authorship, does it not appear plain that all this marvelous growth has some correlation that teaches an important lesson? Why should this mighty wave of civilization set in at such a recent period, and more notably in our own land, when there have been so many nations far in advance of us in point of age? The answer is to be found in the beneficent institutions of our comparatively new and free country, whose laws have been made to justly regard the inventor as a public benefactor, and the wisdom of which policy is demonstrated by the growth of this period, amply proving that invention and civilization stand correlated—invention the cause and civilization the effect.

This retrospect, necessarily cursory and superficial, brings to view sufficient of the great inventions as milestones on the great roadway of progress to inspire us with emotions of wonder and admiration at the resourceful and dominant spirit of man. Delving into the secret recesses of the earth, he has tapped the hidden supplies of Nature's fuel, has invaded her treasure house of gold and silver, robbed Mother Earth of her hoarded stores, and possessed himself of her family record, finding on the pages of geology sixty millions of years existence. Peering into the invisible little world, the infinite secrets of microcosm have yielded their fruitful and potent knowledge of bacteria and cell growth. With telescope and spectroscope he has climbed into limitless space above, and defined the size, distance and constitution of a star millions of miles away. The lightning is made his swift messenger, and thought flashes in submarine depths around the world, the voice travels faster than the wind, dead matter is made to speak, the invisible has been revealed, the powers of Niagara are harnessed to do his will, and all of Nature's forces have been made his constant servants in attendance. We witness a new heaven and a new earth, contemplation of which becomes oppressive with the magnitude and grandeur of the spectacle, and involuntarily we find ourselves asking the question, "Is it all done? Is the work finished? Is the field of invention exhausted?" It does seem that it is quite impossible to again equal the great inventions of this wonderfully prolific epoch; but as these great inventions, which now seem commonplace to us, would have seemed quite impossible to our ancestors, we may indulge the hope of future possibilities beyond any present conception, but onward and upward in the great evolution of human destiny.

Rejoicing in our strength and capabilities, the new light of man's power and destiny breaks more clearly over us, and content with the infinite quality of mind and matter, the teachings of philosophy, and the facts of evolution, we rest in the assurance of positive knowledge that all that has been done in the past is merely preliminary, that human ingenuity knows no limit, and so long as man himself remains hedged about with the limitations of mortality and the conditions of growth, so long will his strivings and attainments be infinite.

INTRODUCTORY NOTE

Henry Adams in his recondite, elusive style, showed his realization of the magnitude of technological power more effectively than did Edward Byrn in his more bluntly stated essay in the *Scientific American.* A visit to the Paris Exposition of 1900 in the company of Samuel Pierpont Langley, Secretary of the Smithsonian Institution, and pioneer in mechanical flight, provoked the thoughts expressed in "The Dynamo and the Virgin." Langley directed Adams' attention to the exhibits showing the energy flow from radium and to the exhibits of the dynamo. Adams found radiation effects elusive and provocative: the energy, or the force, to use his language, was no more and no less real than the radiant energy of the Chartres Virgin. The force of the dynamo was more readily measured and utilized, but to Adams it too was like the Virgin, whose motivating force had built the great medieval cathedral at Chartres. The title of the chapter from which the following selection is taken, "The Dynamo and the Virgin," suggests Adams' concept of man attracted to, and hoping to draw out by supplication, the mysterious and effective power of two sources, the technological and the spiritual. Put more directly, medieval man assumed that the force of the Virgin could radiate from a cathedral; nineteenth-century man assumed that the force of the dynamo could be exploited from a powerhouse. Both dynamo and Virgin drew out the organizing and constructive genuis of man and in turn were expected to respond with even greater and more mysterious power.

Both Adams and Byrn were awestruck in the face of power, but Adams had reservations about the material force of the age of the dynamo, for he wrote that "all the steam in the world could not, like the Virgin, build Chartres," while Byrn in his boundless enthusiasm found no limit as he witnessed, in 1896, the evolution of "a new heaven and a new earth, the contemplation of which becomes oppressive with the magnitude and grandeur of the spectacle. . . ."

Henry Adams' (1838–1918) most widely known work was his *The*

Education of Henry Adams (1907 and 1918), but he enjoys a substantial reputation as an American historian, having written, for example, a monumental study of the administrations of Jefferson and Madison (*History of the United States,* 1889–1891). The *Education* is an erudite and enigmatic, highly eclectic and interpretive autobiographical narrative, social comment, and philosophical discourse. In the concluding chapters, he developed a theory of history strongly influenced by his view of accelerated technological and scientific discovery. He elaborated on this theory in the "Rule of Phase applied to History," written in 1909 (printed in *The Degradation of the Democratic Dogma,* 1919). He was a grandson of John Quincy Adams and a great-grandson of John Adams.

The Dynamo and the Virgin (1900)

HENRY ADAMS

Until the Great Exposition of 1900 closed its doors in November, Adams haunted it, aching to absorb knowledge, and helpless to find it. He would have liked to know how much of it could have been grasped by the best-informed man in the world. While he was thus meditating chaos, Langley came by, and showed it to him. At Langley's behest, the Exhibition dropped its superfluous rags and stripped itself to the skin, for Langley knew what to study, and why, and how; while Adams might as well have stood outside in the night, staring at the Milky Way. Yet Langley said nothing new, and taught nothing that one might not have learned from Lord Bacon, three hundred years before; but though one should have known the "Advancement of Science" as well as one knew the "Comedy of Errors," the literary knowledge counted for nothing until some teacher should show how to apply it. Bacon took a vast deal of trouble in teaching King James I and his subjects, American or other, towards the year 1620, that true science was the development or economy of forces; yet an elderly American in 1900 knew neither the formula nor the forces; or even so much as to say to himself that his historical business in the Exposition concerned only the economies or developments of force since 1893, when he began the study at Chicago.

Nothing in education is so astonishing as the amount of ignorance it accumulates in the form of inert facts. Adams had looked at most of the accumulations of art in the storehouses called Art Museums; yet he did not know how to look at the art exhibits of 1900. He had studied Karl Marx and his doctrines of history with profound attention, yet he could not apply

Henry Adams, *The Education of Henry Adams: An Autobiography* (Boston: Houghton Mifflin, 1918), pp. 379–385, 388–390.

them at Paris. Langley, with the ease of a great master of experiment, threw out of the field every exhibit that did not reveal a new application of force, and naturally threw out, to begin with, almost the whole art exhibit. Equally, he ignored almost the whole industrial exhibit. He led his pupil directly to the forces. His chief interest was in new motors to make his airship feasible, and he taught Adams the astonishing complexities of the new Daimler motor, and of the automobile, which, since 1893, had become a nightmare at a hundred kilometres an hour, almost as destructive as the electric tram which was only ten years older; and threatening to become as terrible as the locomotive steam-engine itself, which was almost exactly Adams's own age.

Then he showed his scholar the great hall of dynamos, and explained how little he knew about electricity or force of any kind, even of his own special sun, which spouted heat in inconceivable volume, but which, as far as he knew, might spout less or more, at any time, for all the certainty he felt in it. To him, the dynamo itself was but an ingenious channel for conveying somewhere the heat latent in a few tons of poor coal hidden in a dirty engine-house carefully kept out of sight; but to Adams the dynamo became a symbol of infinity. As he grew accustomed to the great gallery of machines, he began to feel the forty-foot dynamos as a moral force, much as the early Christians felt the Cross. The planet itself seemed less impressive, in its old-fashioned, deliberate, annual or daily revolution, than this huge wheel, revolving within arm's-length at some vertiginous speed, and barely murmuring—scarcely humming an audible warning to stand a hair's-breadth further for respect of power—while it would not wake the baby lying close against its frame. Before the end, one began to pray to it; inherited instinct taught the natural expression of man before silent and infinite force. Among the thousand symbols of ultimate energy, the dynamo was not so human as some, but it was the most expressive.

Yet the dynamo, next to the steam-engine, was the most familiar of exhibits. For Adams's objects its value lay chiefly in its occult mechanism. Between the dynamo in the gallery of machines and the engine-house outside, the break of continuity amounted to abysmal fracture for a historian's objects. No more relation could he discover between the steam and the electric current than between the Cross and the cathedral. The forces were interchangeable if not reversible, but he could see only an absolute *fiat* in electricity as in faith. Langley could not help him. Indeed, Langley seemed to be worried by the same trouble, for he constantly repeated that the new forces were anarchical, and especially that he was not responsible for the new rays, that were little short of parricidal in their wicked spirit towards science. His own rays, with which he had doubled the solar spectrum, were altogether harmless and beneficent; but Radium denied its God—or, what was to Langley the same thing, denied the truths of his Science. The force was wholly new.

A historian who asked only to learn enough to be as futile as Langley or Kelvin, made rapid progress under this teaching, and mixed himself up in the tangle of ideas until he achieved a sort of Paradise of ignorance vastly consoling to his fatigued senses. He wrapped himself in vibrations and rays which were new, and he would have hugged Marconi and Branly had he met them, as he hugged the dynamo; while he lost his arithmetic in trying to figure out the equation between the discoveries and the economies of force. The economies, like the discoveries, were absolute, supersensual, occult; incapable of expression in horse-power. What mathematical equivalent could he suggest as the value of a Branly coherer? Frozen air, or the electric furnace, had some scale of measurement, no doubt, if somebody could invent a thermometer adequate to the purpose; but X-rays had played no part whatever in man's consciousness, and the atom itself had figured only as a fiction of thought. In these seven years man had translated himself into a new universe which had no common scale of measurement with the old. He had entered a supersensual world, in which he could measure nothing except by chance collisions of movements imperceptible to his senses, perhaps even imperceptible to his instruments, but perceptible to each other, and so to some known ray at the end of the scale. Langley seemed prepared for anything, even for an indeterminable number of universes interfused—physics stark mad in metaphysics.

Historians undertake to arrange sequences—called stories, or histories—assuming in silence a relation of cause and effect. These assumptions, hidden in the depths of dusty libraries, have been astounding, but commonly unconscious and childlike; so much so, that if any captious critic were to drag them to light, historians would probably reply, with one voice, that they had never supposed themselves required to know what they were talking about. Adams, for one, had toiled in vain to find out what he meant. He had even published a dozen volumes of American history for no other purpose than to satisfy himself whether, by the severest process of stating, with the least possible comment, such facts as seemed sure, in such order as seemed rigorously consequent, he could fix for a familiar moment a necessary sequence of human movement. The result had satisfied him as little as at Harvard College. Where he saw sequence, other men saw something quite different, and no one saw the same unit of measure. He cared little about his experiments and less about his statesmen, who seemed to him quite as ignorant as himself and, as a rule, no more honest; but he insisted on a relation of sequence, and if he could not reach it by one method, he would try as many methods as science knew. Satisfied that the sequence of men led to nothing and that the sequence of their society could lead no further, while the mere sequence of time was artificial, and the sequence of thought was chaos, he turned at last to the sequence of force; and thus it happened that, after ten years' pursuit, he found himself lying in the Gallery of Machines at the Great Exposition of 1900, his historical neck broken by the sudden irruption of forces totally new.

Since no one else showed much concern, an elderly person without other cares had no need to betray alarm. The year 1900 was not the first to upset schoolmasters. Copernicus and Galileo had broken many professorial necks about 1600; Columbus had stood the world on its head towards 1500; but the nearest approach to the revolution of 1900 was that of 310, when Constantine set up the Cross. The rays that Langley disowned, as well as those which he fathered, were occult, supersensual, irrational; they were a revelation of mysterious energy like that of the Cross; they were what, in terms of mediæval science, were called immediate modes of the divine substance.

The historian was thus reduced to his last resources. Clearly if he was bound to reduce all these forces to a common value, this common value could have no measure but that of their attraction on his own mind. He must treat them as they had been felt; as convertible, reversible, interchangeable attractions on thought. He made up his mind to venture it; he would risk translating rays into faith. Such a reversible process would vastly amuse a chemist, but the chemist could not deny that he, or some of his fellow physicists, could feel the force of both. When Adams was a boy in Boston, the best chemist in the place had probably never heard of Venus except by way of scandal, or of the Virgin except as idolatry; neither had he heard of dynamos or automobiles or radium; yet his mind was ready to feel the force of all, though the rays were unborn and the women were dead.

Here opened another totally new education, which promised to be by far the most hazardous of all. The knife-edge along which he must crawl, like Sir Lancelot in the twelfth century, divided two kingdoms of force which had nothing in common but attraction. They were as different as a magnet is from gravitation, supposing one knew what a magnet was, or gravitation, or love. The force of the Virgin was still felt at Lourdes, and seemed to be as potent as X-rays; but in America neither Venus nor Virgin ever had value as force—at most as sentiment. No American had ever been truly afraid of either.

This problem in dynamics gravely perplexed an American historian. The Woman had once been supreme; in France she still seemed potent, not merely as a sentiment, but as a force. Why was she unknown in America? For evidently America was ashamed of her, and she was ashamed of herself, otherwise they would not have strewn fig-leaves so profusely all over her. When she was a true force, she was ignorant of fig-leaves, but the monthly-magazine-made American female had not a feature that would have been recognized by Adam. The trait was notorious, and often humorous, but any one brought up among Puritans knew that sex was sin. In any previous age, sex was strength. Neither art nor beauty was needed. Every one, even among Puritans, knew that neither Diana of the Ephesians nor any of the Oriental goddesses was worshipped for her beauty. She was goddess because of her force; she was the animated dynamo; she was reproduction—the greatest and most mysterious of all energies; all she needed was to be fecund. Singularly enough, not one of Adams's many schools of education had ever drawn his attention to the opening lines of Lucretius, though they were perhaps the finest in all Latin literature, where the poet invoked Venus exactly as Dante invoked the Virgin:—

> "Quae quoniam rerum naturam *sola* gubernas."

The Venus of Epicurean philosophy survived in the Virgin of the Schools:—

> "Donna, sei tanto grande, e tanto vali,
> Che qual vuol grazia, e a te non ricorre,
> Sua disianza vuol volar senz' ali."

All this was to American thought as though it had never existed. The true American knew something of the facts, but nothing of the feelings; he read the letter, but he never felt the law. Before this historical chasm, a mind like that of Adams felt itself helpless; he turned from the Virgin to the Dynamo as though he were a Branly coherer. On one side, at the Louvre and at Chartres, as he knew by the record of work actually done and still before his eyes, was the highest energy ever known to man, the creator of four-fifths of his noblest art, exercising vastly more attraction over the human mind than all the steam-engines and dynamos ever dreamed of; and yet this energy was unknown to the American mind. An American Virgin would never dare command; an American Venus would never dare exist.

The question, which to any plain American of the nineteenth century seemed as remote as it did to Adams, drew him almost violently to study, once it was posed; and on this point Langleys were as useless as though they were Herbert Spencers or dynamos. The idea survived only as art. There one turned as naturally as though the artist were himself a woman. Adams began to ponder, asking himself whether he knew of any American artist who had ever insisted on the power of sex, as every classic had always done; but he could think only of Walt Whitman; Bret Harte, as far as the magazines would let him venture; and one or two painters, for the flesh-tones. All the rest had used sex for sentiment, never for force; to them, Eve was a tender flower, and Herodias an unfeminine horror. American art, like the American language and American education, was as far as possible sexless. Society regarded this victory over sex as its greatest triumph, and the historian readily admitted it, since the moral issue, for the moment, did not concern one who was studying the relations of unmoral force. He cared nothing for the sex of the dynamo until he could measure its energy.

[Adams wrote several pages (omitted here) on his friend the sculptor Augustus St. Gaudens. Adams believed that the American sculptor sensed the Virgin of Chartres as a powerful molder of taste but did not see her as a symbol of power. —Ed.]

For a symbol of power St. Gaudens instinctively preferred the horse, as was plain in his horse and Victory of the Sherman monument. Doubtless Sherman also felt it so. The attitude was so American that, for at least forty years, Adams had never realized that any other could be in sound taste. How many years had he taken to admit a notion of what Michael Angelo and Rubens were driving at? He could not say; but he knew that only since 1895 had he begun to feel the Virgin or Venus as force, and not everywhere even so. At Chartres—perhaps at Lourdes—possibly at Cnidos if one could still find there the divinely naked Aphrodite of Praxiteles—but otherwise one must look for force to the goddesses of Indian mythology. The idea died out long ago in the German and English stock. St. Gaudens at Amiens was hardly less sensitive to the force of the female energy than Matthew Arnold at the Grande Chartreuse. Neither of them felt goddesses as power—only as reflected emotion, human expression, beauty, purity, taste, scarcely even as sympathy. They felt a railway train as power; yet they, and all other artists, constantly complained that the power embodied in a railway train could never be embodied in art. All the steam in the world could not, like the Virgin, build Chartres.

Yet in mechanics, whatever the mechanicians might think, both energies acted as interchangeable forces on man, and by action on man all known force may be measured. Indeed, few men of science measured force in any other way. After once admitting that a straight line was the shortest distance between two points, no serious mathematician cared to deny anything that suited his convenience, and rejected no symbol, unproved or unproveable, that helped him to accomplish work. The symbol was force, as a compass-needle or a triangle was force, as the mechanist might prove by losing it, and nothing could be gained by ignoring their value. Symbol or energy, the Virgin had acted as the greatest force the Western world ever felt, and had drawn man's activities to herself more strongly than any other power, natural or supernatural, had ever done; the historian's business was to follow the track of the energy; to find where it came from and where it went to; its complex source and shifting channels; its values, equivalents, conversions. It could scarcely be more complex than radium; it could hardly be deflected, diverted, polarized, absorbed more perplexingly than other radiant matter. Adams knew nothing about any of them, but as a mathematical problem of influence on human progress, though all were occult, all reacted on his mind, and he rather inclined to think the Virgin easiest to handle.

The pursuit turned out to be long and tortuous, leading at last into the vast forests of scholastic science. From Zeno to Descartes, hand in hand with Thomas Aquinas, Montaigne, and Pascal, one stumbled as stupidly as though one were still a German student of 1860. Only with the instinct of despair could one force one's self into this old thicket of ignorance after having been repulsed at a score of entrances more promising and more popular. Thus far, no path had led anywhere, unless perhaps to an exceedingly modest living. Forty-five years of study had proved to be quite futile for the pursuit of power; one controlled no more force in 1900 than 1850, although the amount of force controlled by society had enormously increased. The secret of education still hid itself somewhere behind ignorance, and one fumbled over it as feebly as ever. In such labyrinths, the staff is a force almost more necessary than the legs; the pen becomes a sort of blind-man's dog, to keep him from falling into the gutters. The pen works for itself, and acts like a hand, modelling the plastic material over and over again to the form that suits it best. The form is never arbitrary, but is a sort of growth like crystallization, as any artist knows too well; for often the pencil or pen runs into side-paths and shapelessness, loses its relations, stops or is bogged. Then it has to return on its trail, and recover, if it can, its line of force. The result of a year's work depends more on what is struck out than on what is left in; on the sequence of the main lines of thought, than on their play or variety. Compelled once more to lean heavily on this support, Adams covered more thousands of pages with figures as formal as though they were algebra, laboriously striking out, altering, burning, experimenting, until the year had expired, the Exposition had long been closed, and winter drawing to its end, before he sailed from Cherbourg, on January 19, 1901, for home.

INTRODUCTORY NOTE

Robert Thurston, an engineer and scientist, wrote "The Border-Land of Science" for a broad spectrum of Americans, not solely for like-minded professionals. His enthusiasm for the power and bright future of science and technology emerges unbridled. His word choices set the tone for his essay: "onward and upward movement," "wonderful," "further progress," "acceleration," and "advance . . . never ceasing." With an attitude that might be called "dynamic utopianism," he presumed that men of science, learned engineers, and mechanics would continually add to the sum of human happiness.

Thurston's sounding of the progress theme was by no means fresh, for social philosophers in America and abroad had investigated and celebrated the implications of purposeful evolution. Thurston's emphasis on applied science and technology, however, merits our consideration. "The more we gain, the more is seen to be achievable," he decided. He catalogued the most obvious inventions of the century and told of the "sweetness and light" they brought. He rejoiced because his humbler neighbors had comforts that were once thought luxuries. He also forecast what would be discovered beyond the borderland of science and technology: he foresaw men with more intellectual and women with more pleasing forms; he anticipated an increase of intellectual activity and less back-breaking labor. His extrapolations, in 1890, helped him to forsee airplanes, long-distance power transmission, and the transmission of pictures as well as words. His vision was flawed, however, as he anticipated inventors creating a submarine so terrible that "the death of all naval warfare at a very early date is assured" and in his remarkable prognostication that "the work of the inventor will insure the peace of the world."

Robert Henry Thurston (1839–1903) ranks among the foremost mechanical engineers in American history. His outstanding scientific and engineering investigations were in the properties of materials and in the motive power of steam. His contemporaries, however, knew him best as an engineering educator who planned and administered new courses of study in mechanical engineering at Stevens Institute and Sibley College, Cornell University. His literary abilities and inquiring mind went beyond technical subjects, as the article reprinted here shows. Further evidence of the breadth of his interests are his books, *A History of the Growth of the Steam-Engine* (1878) and *Robert Fulton* (1891).

The Border-Land of Science

R. H. THURSTON

Every intelligent reader and thinker who has studied the history of those marvellous developments of physical science which have characterized the progress of civilization during the last half-century, and who continues to watch the wonderful acceleration of its progress, which seems to-day its most striking feature, must inevitably recognize in its growth an element of vastly greater importance than the merely material and utilitarian side which so disturbed the mind of the eccentric and unfortunate, but always artistic and poetical, Ruskin. Science has been to the world a great comforter, civilizer, and enlightener. It is a moral as well as a physical agent, promoting morality as well as aiding humanity in its physical and intellectual progress. It has generated "sweetness and light" as well as those coarser, but no less essential, elements of our onward and upward movement which seemed to Matthew Arnold so insignificant. Its growth and its effects have been like those of that light described, in Genesis and by Geology, at the beginning of the world's history. At first glimmering, faint, uncertain, hardly visible, exhibiting to the senses only the fact of the existence of infinite darkness, it has gradually spread over the world, growing brighter as it expanded; enlightening wider and wider areas; revealing the good to be sought, the bad to be evaded; its brightness increasing more and more rapidly this side the "dark ages," until we are more than satisfied with the brilliancy of its rays; and as we endeavor to accustom ourselves to the sunlight of our own century, we wonder if it be possible that the race can adapt itself to the requirements of further progress.

But as the light gains in intensity and illuminates more and more brightly the world about us, and as it extends its enlightening rays further and further into the dark regions that always

The North American Review, 150 (1890), 67–79.

border its field, there always exists a border-land, more or less distinctly seen and more or less defined, in which are included those wonders which we may hope yet to see. We continually marvel at the inventions and discoveries of to-day, and wonder still more what will come to-morrow. We are continually expecting to see a limit reached by the inventor and by the discoverer, and are as constantly finding that we are simply on a frontier which is being steadily pushed further and further out into the infinite unknown; and the more we learn and the more we discover, the greater the opportunity for greater and more rapid progress. We are daily learning a more thorough appreciation of our own ignorance and of the insignificance of our finite in the midst of the infinite. We have groped our way with our rushlight, have made better speed with the modern candle, have congratulated ourselves on the excellence of our oil-lamps, have boasted of our gas-lights, and now take exceeding pride in the radiant brightness of our electric lights; yet we are far from an approximation to the volume, the intensity, and the quality of sunlight, and can see that the path ahead is still too long for our measurement, and that, in comparison, the distance already traversed is microscopic. The border-land is still ahead of us, constantly enlarging as we move on. The more we gain, the more is seen to be achievable.

The progress observed by the student of history from rushlight to electric light, and from barbarism to civilization, has always been an accelerated motion. From insignificant beginnings we have seen the race advance by a movement, slow at first, gradually increasing, continually gaining in rate of motion as well as in position, until, especially during the last generation, the velocity of onward movement has been such that the brightest intellect, the most powerful mind, is utterly unable to follow it in all its paths, and every worker and every student has become a specialist. Each is well content to contribute his mite to the general treasury and to assist by aiding ever so little through his labors in his own restricted field. No Humboldt can ever again grasp the whole of existing pantology; no Bacon can ever again hope to see the limits of the ever-widening field; no Compte can ever again safely attempt to plan a scheme for the development of all minor lines of scientific investigation.

Excursions into the border-land of science must hereafter

be planned and conducted by men who have already made themselves familiar with the previously-explored adjacent territory; and we find that the accelerated progress of the past and of the present is likely to be assured by a greater and greater body of such specialists. A hundred philosophers follow Bacon; a thousand seek the ends proposed to them by Compte; and innumerable scientific students and investigators endeavor to make known the unknown by entering upon the paths of which the beginnings were revealed by earlier Humboldts. A few years ago Faraday studied chemistry and physics; to-day he would be either a very stupid or a very rash man who should claim to be at once chemist and physicist. In the last generation we had chemists and physicists; to-day the chemist finds ample scope for all his powers in the study of the molecular structure of an organic compound, or in the investigation of the petroleums, or in the examination of the effects of varying proportions of manganese upon the properties of the modern "mild" steels; while the physicist has become an electrician, a student of the spectra, an investigator of the form of the sound-wave; or he studies the conditions affecting the liquefiability of the so-called "permanent" gases. Thus specializing, he becomes competent to attempt researches in the border-land of his science and to attack problems now vaguely looming up ahead like great icebergs, far away on the distant sea, obscured by their own mists—problems which are even grander and more important than those which the nineteenth century has already seen solved.

Thus, as a glance at the history of the past and its progress toward the present readily shows, the movement of the great current, while always more or less variable in rate and in direction, under the irregular action of impressed forces, has always, on the whole, been onward, and with, on the whole, an accelerated motion. The forces which have acted to modify, to change, to retard, or to accelerate this progress have been as apparently variable as those of the winds and the tides; but, like those of wind and tide, every force so acting, whether moral, physical, or intellectual, has been controlled by law, and the resultant effect has always and invariably been an onward motion, with continual acceleration.

Looking back, then, upon this past of the race, we have seen mankind emerge from barbarism, become civilized and enlightened, growing in every possible direction of application of his faculties; passing through a long period of slowly-progressing advancement with still less easily perceived acceleration; gaining a hold upon one after another of the sciences and the arts; securing, one by one, the means of self-preservation and self-aid; all through the middle ages working for safe and stable forms of government; finally, after a certain stability had been reached, entering upon an era of thought and scientific investigation that was like the action of a new impulsive force driving the peoples of Europe along their rising pathway toward higher and greater life, with a now more and more perceptible gain in rate of motion. A century ago, the first fruits of the first efforts of scientific investigators of the preceding century began to be seen; and since the days of Lord Bacon and of Shakespeare, of Milton, and of Boyle and Young, into and through the times of Lavoisier, of Humphrey Davy, of Rumford, and of Faraday and Watt, of Stephenson, Fulton, Morse, and Wheatstone, of Stevens and Evans, and of the wonderful train of discoverers in science and of inventors in the arts who accompanied and followed those pioneers in exploration of all the innumerable fields opened by them in the last and in the present century, this acceleration has been more observable than ever before, and its effects have been more than ever impressive in their magnitude and in their results in the promotion of the welfare of humanity.

The centenarians of to-day—and they are more numerous than we commonly think—have seen the growth of nearly all that makes modern life. They saw the steam-engine introduced and applied to the turning of the busy wheels of countless mills, producing all the fabrics and all the apparatus essential to our life and comfort. They saw the railway laid in lines of iron and steel across continents, and forming bonds of far more strength than any treaties to hold State to State and to preserve the nation and the liberties for which our fathers and our grandfathers fought and died. They saw the inception and all the growth of steam navigation, and the construction of ships of continually increasing magnitude and speed, crossing the oceans like mighty shuttles, weaving the web of union between countries separated by thousands of miles of seas, and netting all nations in a community of interests that shall, in good time, become the grandest and most efficient safeguard against internecine feuds, and the most effective security against that most terrible and most shameful of all "relics of barbarism"—War. They have seen the lightning captured and harnessed for the noblest purposes of life, giving man conference with man across continents and seas, carrying his messages under the Atlantic, over all Europe and Asia, among the islands of the Indian and Pacific oceans, and under thousands of miles of water and over thousands of miles of land to farthest Australia. They have seen all energies converted to the use of mankind. They have seen the stored heat-energy of the rays of the sun, converted, as they were, millions of years before man came on this planet in visible form, reconverted into power and applied by the steam-engine to the production of electricity, and this new force set at work to transmit power over the world, to give us light in our streets and dwellings, and to the production of new forces in endless ways.

On these great forces which are the life of industry, the vital forces of steam and of electricity, are resting all the social edifices of modern life. All that we eat, all that we wear, our houses, every comfort, and all our luxuries are brought us to-day by these invisible, but almost omnipotent, genii of the fire and the lamp and the thunder-storm. The steamboat, the railroad, the telegraph, the telephone, the modern printing-press, all our machinery of the arts, and every recent invention and applied scientific principle, are brought us or are worked for us by them. The result is that our humblest neighbors, having health, and with habits of industry and wise forethought, may readily gain such comforts, and such once so-called luxuries, as no royal family could boast in the days of our grandparents—comforts and luxuries that were not then in existence and which no wealth could then procure. A nation may now become educated; a people may now be safe against poverty or famine; the world is even now, probably, past the critical point and sure of unintermitted future progress.

What more can we ask? What more may we expect? What more have we any right to hope for? Is this advance to be never-ceasing? What does science and what does the judgment of wise men justify us in hoping for? What are the discoveries and the inventions that science and art may be expected to give us in the future? These questions cannot, as a matter of course, be fully or definitely answered; but it is, perhaps, possible to obtain some idea of the extent and direction of this motion of the life of Man, this approximately, if not accurately, straight line of progress, for a little way ahead. We may, at least, ask what is the path now open, and what may we hope to find as we advance a little way further along it. Perhaps we may even be allowed to indulge in some speculation, if we carefully endeavor to distinguish between the results of our "scientific use of the imagination" and the reasonable deductions of permissible argumentation. We will take our speculations first and our more thoroughly founded deductions afterwards.

I am always inclined to ask, first, what may we believe to be the probable form and likeness of the coming man and his wife. I imagine that, when we look back from our home in the unseen universe, ages hence, we shall see, without much doubt, a race of men differing from those of to-day much as the man of to-day differs from his simious, perhaps simian, ancestors. The brain will be developed to meet the more complex and serious taxation of a more complex and trying civilization; the vital powers will be intensified; the man, reducing the powers of Nature still more completely to his service, will depend less on the exertions of his muscles, and they will be correspondingly and comparatively less powerful, though they will probably, nevertheless, I imagine, continue to grow somewhat in size, as they unquestionably have grown since the middle ages; the lungs must supply aëration to a larger and more rapidly circulated volume of blood richer in the phosphatic elements especially needed for the building-up of brain and nerve; the digestion must supply its nutriment in similarly increased amount and altered character and composition; the whole system must be capable of more rapid, more thorough, and more manageable conversion of the energies of the natural forces to the uses of the intellect and the soul which inhabits it.

If so much be granted, it is easy to see something of the nature of the change in the physical man that must gradually take place. The brain will enlarge in its anterior even more than in its posterior parts, and the great forehead will probably overhang a heavy but mobile face, having a god-like intelligence of countenance; with eyes large and prominent; with large nostrils; with a set of jaws at once fitted for the reduction of grain foods to pulp and to give basis for muscles capable of expressing great ideas by word and by play of feature. The chest will be large; the lungs capacious and free in operation, promptly self-adjusting to all demands and all variations of demand, and fitted to aërate enormous volumes of fluid flowing in from the veins. The digestive organs will necessarily be suited to develop and apply the phosphatic nutriment of grain and fruit foods; the liver, and spleen especially, producing those fats which make the main part of brain and nerve tissue—the abdomen thus growing with the lungs. The limbs may probably be longer; better cushioned with fat than now; smaller in proportion to the rest of the body, as to weight at least; though we may presume that this change will be made with positive gain, on the whole, in grace and general power. A more generally intelligent race will pay more attention to the preservation and cultivation of the physical powers by exercise and every sanitary device, and this will unquestionably aid in the development of a noble physique. The coming man will be tall, and free and lofty of carriage, as will befit a being full of noble thoughts and high aspirations, and his progress toward the infinite in all that is good and great will be commensurate with his ennobled powers of body and mind.

The woman of the coming race will have a similar development. Mind and body altering in similar directions, her intellectual face and her noble head will be carried above a no less impressive form. Ages of further growth of her always-controlling affections will have conferred upon her, even more than upon her consort, those beautiful perfections of manner and those attractions of face and figure, coming of the freer and freer play of the affections and graces of home, which must always distinguish in a superlative degree the lovelier sex. She will grow with the ages and through the ages; her form will gain in grace and strength, in roundness and beauty; and she will, as always, lead man in his approach toward heaven. Easier lives and more intelligence, a better application of knowledge, and in all respects a better life, will give to both the most that nature can confer. As evil dies and virtue survives and strengthens, they will, hand in hand, advance continually, and with continually greater ease and speed, toward the perfect life.

"What may we hope from science?" No man can say; but perhaps vastly more than the most sanguine to-day venture to predict. We have seen steam applied to the propulsion of the steamship and the railway train. Within a few years the steamboat has grown from the dimensions of a yacht to the size of the "Great Eastern" and of the "Sardegna" and the "City of Paris," driven by the power of 15,000 to 20,000 horses, crossing the Atlantic in less than a week. Another generation or two may see the size doubled and speed still further increased, the voyage reduced to less than four days. What all this means it is difficult to conceive; but perhaps it may aid the imagination to say that the engineer's horse-power is about a half more than the actual power of the average horse; that the engineer's horse works unremittingly twenty-four hours a day, and thus does the work of three real horses; that this means that he must find a way to stow the equivalent power of about a quarter of a million of horses in his ship of the next generation; that he must put the power of nearly 150,000 tons of horses into 8,000 tons of engine; that he must snugly pack in a ship 1,000 feet long, or perhaps less, the power of a "string-team" 500 miles long. But all this seems perfectly possible, and not altogether improbable, to the hopeful engineer of to-day. He thinks it may prove to be only a matter of time and money.

On land, steam is likely yet to show powers that may astonish the spectator of its performances as much as at sea. Speeds of sixty miles an hour and over have been already frequently attained, and there seems no reason to doubt that this is but a beginning. Boys living to-day may very probably see speeds exceeding a hundred miles an hour attained, and the continent crossed in two days or less. In fact, I have reason to believe that a speed of ninety miles an hour has already been maintained by an ordinary engine and light train for a short distance, with one of my over-bold boys at the throttle-valve; and it may not be long before engines are built for such speeds. Such an engine leaving New York in the morning would reach San Francisco the next night. But before this can be done much work will be required on the road-bed, as well as in the improvement of the machine itself. The engineer, however, given the demand and the capital, can unquestionably find ways of securing such speeds and maintaining them continuously and with safety.

But both these advances may become commonplace beside other wonders that we may reasonably believe possible to the wonder-working mind and hand of man, to the inventor and the mechanic aided by all the resources of modern and future sciences.

We are already familiar with the telegraph, transmitting messages across land and under sea around the world; we have even discovered a method of sending over its wires the *facsimile* of a written, or a printed, or an engraved page; the hand which writes a letter at one end of the line is imitated at the other by the pen of the lightning, and his every line, curve, and dot and dash exactly reproduced under the eye of his waiting correspondent. Who knows but that the time may come when his portrait may go with his letter, or even the words of his mouth, sent through line of telephone, be apparently the issue of the familiar face speaking, like a voice from the Arabian Nights, across the world? The Morses, the Bells, and the Edisons of coming years may be relied on to perform no less wonders than those which now astonish us; and the phonographs and graphophones of later days and of future ages will record in their own voices and their own language, for all the eternities of earth-existence, their thoughts and their triumphs, in dormant, but ever-living, form. The great electricians of to-day are teaching us how to convert the energy of the steam-engine into the newer form and to apply it to the illumination of our streets and our dwellings, and to employ it in the operation of all the machinery of shop and mill and home. The time may yet come when, by the employment of this wonderful conveyer of power, the energy of all the coal-mines, or of the immeasurable tons and tons of water pouring over that tremendous precipice of Niagara into the abysses of the whirlpool and the rushing rapids below—over three millions of horse-power—may be transmitted along a copper wire to distant cities to furnish the motive power of factories, of workshops, and of innumerable home industries, doing its share of the great work yet to be performed, of breaking up the present factory system and enabling the home-worker once more to compete on living terms with great aggregations of capital in unscrupulous hands. Great steam-engines will undoubtedly become generally the sources of power in our larger cities, and will send out over the electric wire, into every corner of the town, their Briarean arms, helping the sewing woman at her machine, the weaver at his pattern loom, the mechanic at his engine lathe, giving every house the mechanical aids needed in the kitchen, the laundry, the elevator, and, at the same time, giving light, and possibly even heat, in liberal quantity and intensity. It may become a more powerful genius than was ever dreamed of by Scheherazade, and described in the "Arabian Nights' Entertainments."

There are some other directions in which we may certainly hope to see as marvellous changes in the future as have been witnessed in the past. It is not impossible that we may see fleets of submarine boats doing the work of peace and of war. A century ago, nearly, Robert Fulton, following in the path opened by Bushnell still earlier, built in France a boat which, sailing about on the surface like any other craft, would then strike its sails and plunge beneath the surface, moving about at the will of its commander by the hour, would reappear where least expected, shake off the floods from its decks, and, raising its mast again, steadily sail across the sea to its destination. In our own days, my friend Holland, the most persistent, the bravest, and most reliable of inventors in this art, has built his submarine boats to carry himself and friends about the harbor of New York, spending the hours under water or on the surface, as he might choose, and showing his turtle-shaped deck at one time off the docks of the city, and, an hour or two later, astonishing and frightening the passengers of a steamer in the Lower Bay by his sudden rise alongside. Such boats will probably be used in submarine explorations, and will undoubtedly be employed in naval warfare, to the confusion of nations spending their millions upon the monster iron-clads now familiar to us. The limited experiences of our people during our own Civil War, with their rude "davids" of that time, showed what may be anticipated when these submarine craft are made capable of life at sea and of traversing long routes. Possibly we even may hope that the time will come when we, with some later Captain Nemo, may thus cross the Atlantic, unaffected by gale or wave, in comfort and safety, winter and summer alike. The problem is unquestionably in a promising state of semi-solution. When the submarine boat, the Howell or other torpedo, and the Zalinski gun are brought together in one such craft, the death of all naval warfare at a very early date is assured. The work of the inventor will insure the peace of the world.

Storage batteries, with their stored energy in the form of electricity, may be useful in submarine navigation and in the propulsion of carriages on land; and it looks as if they were likely at an early date to be of service in the solving of that greatest of all the visible problems of the engineer, the navigation of the air. For this purpose it seems possible that steam, too, may yet be of service, to at least a limited extent. It was shown, many years ago, by the distinguished French engineer Dupuy de Lôme that balloons might be made "dirigeable" by hand and impelled by power, and his work has been repeated in later years by the Messr. Renard and Krebs, using stored electricity and attaining a speed of nearly fifteen miles an hour. An English engineer also, Mr. Pole, the veteran authority on applications of steam power, asserts that, taking the lifting power of the balloons as given by latest good work, and assuming that they may be propelled by engines no heavier than those sometimes employed in torpedo-boats, it would be possible to drive a cigar-shaped balloon, four hundred feet long and thirty feet in diameter, at the rate of thirty miles an hour. That this will be realized, and more, we may hardly doubt. But it remains a question whether the balloon can be dispensed with and the flying-machine made self-sustaining as well as self-impelling. This is greatly doubted by many; but it is impossible to say positively that it cannot be done.* We may possibly yet see the air navigated by flying-machines of enormous size, conveying passengers and important despatches, and perhaps light articles of merchandise, at rates of speed exceeding the flight of the swiftest birds; but few men of science or engineers imagine that such machines will ever become vehicles of general use, or as reliable as to time and dates as are the steamships or the railway trains of to-day. Should the inventor of the successful flying-machine ever come forward, he will meet with a welcome such as has rarely been accorded any one of his predecessors. He will have performed a more wonderful task than any one of them.

A still more wonderful work will be done by the genius, should he ever appear and should the thing be possible, who

*Recent experimental investigations by Professor Langley, the distinguished astronomer and physicist, prove that the difficulties here met with are vastly less than had been previously supposed.

shall find a way of producing that beautiful and incomprehensible light emitted by the fire-fly or the glow-worm—a light which is without heat and illustrates probably the only known case of at least approximately complete transformation of vital or heat-energy into light without waste and at low temperature. Such a transformation occurs less perfectly in phosphorous combustion, and the hint given by the animal and by the mineral cannot be supposed to be quite beyond the limit of thought of coming discoverers and inventors. He who turns the glow-worm's light to the use and benefit of mankind will confer an inestimable boon upon the race. It seems perfectly reasonable to suppose that this, at least a problem already solved by Nature, is not entirely beyond the reach of the ever-fruitful mind of man. It would seem no more improbable that the chemist should detect the secret of the composition of the fire-fly's illuminant than it once appeared that he should ever effect the synthesis of madder and revolutionize a great industry. Not only would the solution of this problem be a benefit to the race, as giving them a most beautiful and mild light, but the conversion of heat or other energy into this form of light, without wastes, would result in the most extraordinary economy. It is estimated that, in the case of the more common forms of light-producing apparatus, not over 5 per cent. of the energy is made useful, the remaining 95 per cent. being wasted as heat, and worse than wasted, as the heat produced is always a source of annoyance and loss of health. He who shall give us the secret of the flashing out-of-door lights of the summer evening will enable us to secure twenty times as much light with a given expenditure of fuel as we now obtain, and will in that proportion both cheapen the production of light and reduce the amount of injury done by combustion. We boast to-day of our electric lights; but, this invention or discovery made, we shall have a far less expensive, though perhaps not more healthful, light.

The direct conversion of heat into electricity, or the direct production of that fluid by the combustion of fuel, is another of those problems which are thought by many men of science to be possibly capable of solution, and some rather promising efforts have been made to reach a result so attractive. This may even prove to be the true solution of the problem of the glow-worm. Should the time ever come when, by the burning of a little coal in our houses, we may at once heat them comfortably, derive all the power needed to do the domestic drudgery, secure a beautiful moon-like light, and preserve them, at the same time, against sensible vitiation of the atmosphere, we shall have attained a state of beatitude which might well be taken as a foretaste of paradise. We can, and actually do, transform heat and other energies already within our own bodies at a constant temperature: why may we not repeat these processes outside of them? It does not seem possible that we must always submit to the now inevitable thermal and thermo-dynamic waste of three-quarters to nine-tenths, or more, of the stored energy of our fuels.

We can, at best, only speculate about these coming blessings of a future time and the coming race; but we may, at least, be permitted to hope that much of this possible may become actual. We may be allowed to hope that later generations may continue to see an interminable succession of advances made by coming men of science, and by learned engineers and mechanics, that shall continually add to the sum of human happiness in this world, and make it continually easier to prepare for a better world and a brighter. Who knows but that the telescope, the spectroscope, and other as yet uninvented instruments may aid us in this by revealing the secrets of other and more perfect lives, in other and more advanced worlds than ours, despite the head-shaking of those who know most of the probabilities? Who can say that the life of the race may not be made in a few generations, by this ever-accelerating progress of which the century has seen but the beginning, a true millennial introduction into the unseen universe and the glorious life that every man, Christian or sceptic, optimist or pessimist, would gladly hope for and believe possible? Of this we may be certain: no one can imagine the reach and limit of the results of the application of the intellect of man to the problems of life to-day, any more than could our ancestors of two centuries ago have imagined or believed in the progress that we to-day may look back upon. The border-land of science still stretches on into the unknown.

INTRODUCTORY NOTE

In the opinion of many Americans, Thomas Alva Edison best represents American inventive genuis. He was the hero of America's era of technological preeminence. Plain-spoken, practical-minded, and believed fantastically successful, he was considered in his later years a wise man and venerable prophet whose views on American policy and society should be sought. It is not surprising, then, that *The New York Times* featured an interview with him in 1915 in which he announced his plan for national preparedness (invincibility without overburdening taxes).

Obvious in his remarks was his assumption—like that of so many Americans—that America could solve old problems, like war and preparedness, in a new way that involved technology. Technology was to him and to them the near-omnipotent problem-solver, the moving force of the modern era. With regard to the problems of warfare, he asserted, "Modern warfare is more a matter of machines than of men. Most of the machines are simple matters if we compare them to the machines of industry."

In essence, Edison recommended that America draw up blueprints, build master models, and construct pilot plants that could quickly and easily be expanded in case of war. A small standing army and navy would be staffed by officers and men who in peacetime would return to industry, for it was there that the advancing front of American technology was located. The industrial experience would thus be diffused to the army and the navy. What proved to be his most effective long-range proposal was that the government should establish a great industrial laboratory in which "could be developed the continually increasing possibilities of great guns, the minutiae of new explosives, all the techniques of military and naval progression, without any vast expense." As a precedent, he had in mind his own research laboratories at Menlo Park and West Orange, New Jersey, which have been called the first invention factories—places where invention became a method. His recommendations contributed substantially to the founding of the Naval Research Laboratory, but Edison was disappointed because the laboratory did not fulfill his expectations.

As the subsequent selection about Josephus Daniels and the Naval Consulting Board shows, Edison's views and his willingness to serve in the cause of national preparedness led directly to the formation of the Naval Consulting Board. Its formation was a corollary of the American conviction that inventors were a national resource in wartime as well as in peace.

(*Davis*) *Edward Marshall* (1869–1933), born in New York State, was a foreign correspondent, feature writer, newspaper editor, and author of books and plays. From 1910 to 1914 he wrote interviews for *The New York Times* and Sunday newspapers throughout the country. His career involved him in many dramatic events and with the leading personalities of his time. Among his books is *Story of the Rough Riders,* 1898.

Edison's Plan for Preparedness

EDWARD MARSHALL

The inventor tells how we could be made invincible in war without overburdening ourselves with taxation.

Thomas A. Edison has been considering the relations of the United States to the European war and the possibilities that we may some time be involved in a great conflict. The great inventor is no peace-at-any-price man. His career has shown him to be a fighter. But he is not a militarist.

He believes that we should be invincible. In the following interview he for the first time tells the world how he thinks we may accomplish this without so burdening ourselves with taxation as to reduce our living standards and morale to the European level.

His plan for rendering us invulnerable to attack, while at the same time preserving us from high taxation, includes the establishment of new West Points and new Naval Academies for the training of officers and a vast system of military and naval education for the rank and file.

He would establish vast reserves of stores and arms and ammunition and he would count rather upon automobiles than upon the railroads for quick transportation.

He would build many aeroplanes and submarines, and he would construct a fleet of cruisers, battleships and other naval vessels—this is his most extraordinary proposal—to be kept in drydock, practically in storage, and fully up to date, until needed.

We discussed the matter while we sat in the great library of his laboratory at Orange.

"Several things already have been proved by the war," said Mr. Edison. "One, of course, is that war itself is inefficient. But

---

The New York Times Magazine, May 30, 1915, pp. 6–7.

we knew that. Another is that an 'efficiency' which submerges the individual is an inefficiency.

"As a lay student of the situation it seems to me that the comparatively untrained Englishman has had an advantage from the start just because he has been untrained. This is a striking thing, with a big lesson in it, for the English soldier, I believe, may be regarded, upon the whole, as the physical inferior of the German soldier. Too much military training not only availed Germany nothing, but actually proved to be her handicap.

"Germany was ready for war after the old idea of readiness, but her army never got to Paris. She was overready. She was so overready that she was nervous. Her trigger-fingers became jumpy. It was an attack of hysteria, due to overreadiness, which plunged Europe into war.

"Another thing which has been proved is that no engine of destruction or defense can be so effective that the ingenuity of desperate men cannot devise something which will offset it. Germany's new field guns, the secret of which had been so carefully kept, were the sensation of the first weeks of the war, yet France matched them before it was too late.

"In the unavoidable interpretation which one must place upon these facts is another reassurance for America. We are as clever at mechanics, whether they be those of war or those of peace, as any people of the world. We gave the world the ironclad war vessel as the result of one emergency. We gave the world the submarine. Our Wright brothers perfected the aeroplane.

"If any foreign power should seriously consider an attack upon this country a hundred men of special training quickly would be at work here upon new means of repelling the invaders. I would be at it, myself. There would be no lack of the spirit of determination or the spirit of self-sacrifice. Of these two qualities was the 'Spirit of '76' made up. It is still latent here.

"I believe that the developments of the European war have proved beyond the shadow of a doubt the uselessness of large standing armies. The best work which has been done has been that of the English and French volunteers and the German landsturm.

"It has been a war of trench fighting. What does all the

elaborate training of manoeuvres count in trench work? And what does the fact that it counts little mean? Certainly that the world has wasted a vast amount of money in unnecessary military drill and useless fortifications. I cannot understand the situation in any other way.

"I do not wish any of these statements to lead readers to the belief that I would have my country neglect to realize the necessity of being able to defend itself. I merely wish to call attention to the lessons which the European war seems to me to teach.

"I consider it a reasonable certainty that some day we shall have a war; and I consider it a probability that when that day comes we shall find ourselves unprepared to meet it. I believe it to be the duty of every American patriot to do what he can to see that this does not occur, but I do not believe that the events of recent months in Europe have shown their method of preparation to be the right one.

"Always we have done new things or done old things in a new way, and frequently they have been better things and better ways than Europe has developed. Why should we follow her lead in a military course which has proved to be disastrous to her?

"The European plan of readiness for war really has provoked war. We should evolve a plan of readiness for war which would not do that, but which none the less would worthily protect us.

"We should not take our men from industry and overtrain them, but we should have 2,000,000 rifles ready, in perfect order, even greased, with armories equipped with the very best machinery to begin upon short notice, in case the work should be required, the manufacture of a hundred thousand new firearms every day.

"We should not only have upon hand a large surplus stock of the best ammunition, but we should have Government factories equipped to produce a thousand tons of high explosive in a month if need arises.

"We should have a thousand trenching engines ready and should be prepared with every other mechanical device for rapid defense. Of these things I am certain.

"But I do not in the least agree with the advocates of a great standing army or even of a great military reserve. I believe that, all other details having been looked after, we shall be quite safe if we maintain, as now is authorized, an army of, say, 100,000 men.

"With as many men as that with which to meet the first shock of an emergency, I believe that we could confidently count on volunteers to meet what might come later.

"We should organize our State militia upon really efficient lines. It is my belief that it should be under national, not State control. The men who train it, whether their selection be left with the States or be the business of the National Government, should be chosen with as much care as that with which I select men for important tasks in my laboratory.

"The development of such a method quickly would discover for us, in addition to our standing army, at least 25,000 men especially equipped by natural ability and taste to achieve military efficiency, and these would be drill sergeants, competent to instruct quickly a vast number of soldiers in time of emergency.

"I have suggested 25,000 drill sergeants. We would be doing better if we had 40,000.

"What we want is a small army trained to a big knowledge, and trained to teach it as well as to exercise it. Raw material for training is at hand. We have many millions of potential fighting men.

"We never must become a military nation in the old sense of the term, but I believe it possible that we may become one of the greatest of the military nations without burdening ourselves with any comparatively great, permanent military expense. Modern warware is more a matter of machines than of men. Most of the machines are simple matters if we compare them to the machines of industry.

"If we had machinery at hand with which to equip a million men we could find the million men upon twenty-four hours' notice. There is practically no military sentiment in the United States, nor ever has been, but we have proved ourselves to be among the world's most powerful fighters whenever we have had to fight.

"What is true of our necessities for machinery is true, also, of our necessities for a great supply of field pieces, large cannon, and ammunition. We should have a large number of small factories, equipped and with the raw material at hand in quantities, but so stored as to avoid deterioration, ready to make the latest and most powerful explosives. We should have arsenals with an enormous capacity for the manufacture of large guns, and their facilities should be kept strictly up to date; we should have accurate knowledge of all shops and factories equipped to manufacture tools for defense, aeroplanes, and all manner of accoutrements. We should have contracts with the owners permitting the commandeering of all such shops in case of war and at a given price for their use, and this should be true of all instrumentalities needed in case of war and instantly operated.

"We need not keep men employed in these shops out of more productive work in times of peace in order that they may be ready to give service if a war should come.

"We should carefully consider transportation in its changed conditions. The efficiency of the railroad is not, now, a matter of such vital moment for us as a means of moving troops, although, of course, the railroads must remain for many years the chief means by which heavy artillery and supplies will be moved.

"The motor car is more flexible than the railroad, and our roads are reaching such a stage of betterment that automobiles could be generally utilized for moving men.

"Of course, in case of war, troops would be needed on the coasts. Less than 5 per cent. of our country would need defense. All our war would be there.

"The greater part of the transportation to the Eastern coast could be most efficiently done by automobiles.

"I do not believe we would lack transport if we organized an emergency system by means of which our vast number of privately owned motor cars could be commandeered in case war came. It would be easy to commandeer 200,000 automobiles, and 1,000,000 men could be moved 100 miles in a night by using all the parallel common roads.

"I am the last man who would be willing to suggest parsimony in expenditure upon coast and harbor defense. We should have more guns than we have now at all our harbors and they should be better guns, of longer range than any ship can carry.

"That ought not to be a difficult problem to work out, when it is considered that the harbor defense guns would be mounted upon solid foundations, while ships' guns must be mounted upon platforms of a limited carrying capacity.

"I advocate not only the construction of an enormous number of submarines, as I have suggested, to be held in readiness for operations, not to be kept in commission, but our manufacture at once of a vast supply of harbor defense mines and the construction of many vessels properly equipped to plant them hurriedly in case of an emergency.

"In trench fighting, with our unlimited supply of the most intelligent and independently thinking individual fighters in the world, we would be invincible. In case we were attacked we could set our theatre of defense to suit ourselves, planning (these figures are wholly tentative) fifty lines of trenches.

"The first line, or even the first two or three lines, would be dug as practically all those in this European war have been dug, by individual soldiers with picks and shovels, but lines to the rear of them could be dug (and this is one of the emergencies for which we should prepare) by trenching machines. We have developed this line of machinery to a state of very high perfection and to adapt the existing machinery to the purposes of military trenching would be a very simple matter.

"With fifty or more lines of trenches thus quickly, perfectly, and very cheaply prepared we could easily defeat, even destroy, any attacking force which the enemy might land from his ships.

"He probably would be able to take some of our first lines of trenches, but it is inconceivable that he could have any men left with which to fight after he had reached, for instance, (to select a numeral at random), our twenty-fifth line.

"If the veritable worthlessness of great standing armies and the wicked waste of their maintenance may be considered the most important lesson which the European war so far has held for us, the value of the simple, inexpensive trench is next in importance to us.

"Europe has been conducting a vast and terribly costly experiment for our benefit. She has shown us that in thirty days we can organize a more effective army than the Germans have been able to put into the field if we follow, with the rank and file, the plan of preparation which I have suggested, giving the men the rudiments of training and then returning them to industry.

"She has shown us that we need trained officers. We should immensely increase facilities for training them, even to the establishment of many schools as efficient as West Point.

"But these men, too, should be returned to civil life, after they have had their training, with annual periods of additional study to keep them up to date. They should not be taken permanently from productive and thrust into unproductive effort. They should be kept alive, alert, abreast of everything worth while; we should make splendid all-around citizens of them, fit for unusual usefulness in civil as well as in military effort.

"I think we never should let up on training men for the navy. We should have the greatest number of trained naval sailors that any nation ever has had, but we should not let them eat their heads off after they have got their training.

"We should greatly increase our number of competent naval officers, but we should not make the work of most of them a life career. Like the officers we train for military service, our naval officers should be developed to the top notch of efficiency and then sent back to private life upon part salaries and required to keep up with new developments and be ready for a call if one should come.

"I believe that we should have a navy larger than our present fleet, probably much larger, but I do not believe that the additional ships should be kept in commission.

"I should not in the least object to the payment of my share of the tax which would be necessary for the construction of a dozen dreadnoughts or, for that matter, of two dozen dreadnoughts, but I should strenuously object to the payment of a tax for the support of all of them, manned and in commission during days of peace.

"After each ship is built it should be launched and tested, and then, like the arms and ammunition, it should be stored till the day of need came. Enough vessels of the most approved type should be kept in commission to be used as training ships and enough men should be trained so that we would have no difficulty in finding competent crews for all our vessels. Create a great surplus of trained men, then send them back to industry, with payment of a small annual retainer.

"I believe that in addition to this the Government should maintain a great research laboratory, jointly under military and naval and civilian control. In this could be developed the continually increasing possibilities of great guns, the minutiae of new explosives, all the technique of military and naval progression, without any vast expense.

"When the time came, if it ever did, we could take advantage of the knowledge gained through this research work and quickly manufacture in large quantities the very latest and most efficient instruments of warfare.

"England is doing great work, now, with wonderful artillery. By far the greater part of these big guns have been created out of raw material since the beginning of the war. They seem to be as effective as if not more so than the German guns, which were made in advance of and in anticipation of the conflict, succeeding many other guns, made in former years of peace, but, becoming antiquated, presumably melted up to furnish some of the material for the new artillery.

"At this great laboratory we should keep abreast with every advanced thought in armament, in sanitation, in transportation, in communication—as, for example, under the last named head, with the rapidly developing telegraph and telephone, and, under the head of transportation, with motor car building.

"If we did this we very quickly could manufacture supplies in wholesale quantities when the need for them arose. We could see to it that no attacking nation could have longer-range or more accurate artillery than we would be prepared to make upon short notice."

I asked Mr. Edison to specifically comment on the movement which formed one of the most conspicuous features of the recent Congress—that led by Mr. Gardner for a vast increase in our military expenditure.

"The Gardner movement is unqualifiedly bad," he answered without hesitation. "We don't need any such preparedness as he and his associates are advocating. For General Leonard Wood I have the highest and most profound respect; but I do not agree with him in his opinion as to what is necessary to the welfare of this country in the way of a military establishment.

"We do not need the great machines which these undoubtedly well-intentioned gentlemen are advocating. There is infinitely less reason to believe, today, that we need them, than there was before the outbreak of the European war. We now know how to fight. We did not know, Europe did not know, until this war developed."

INTRODUCTORY NOTE

By August of 1915, with the war raging in Europe, Americans, especially Navy Secretary Josephus Daniels, were concerned about the state of American defenses. At the time, battleships and submarines were perhaps the most complex of military weapons, so Daniels and American naval officers were determined to have advanced technology utilized by America. This concern was heightened by the realization that because of war-imposed secrecy, America could not adopt and adapt European innovations. Critics of American naval technology pointed out that much of it had been borrowed ("Chinese copying," it was then called). Daniels' reaction to this concern was indicative of an attitude current at the time: America's unique inventive genius was a power that could be brought to bear to solve her most pressing problems. Adams and Byrn had written of invention and technology as if they were abstract forces that could be directed at will to an unending variety of purposes, so it is not surprising that Daniels and others turned to this resource in time of war.

Following an indication from Thomas Edison that he would respond to America's need for stronger defenses, Daniels decided to establish a board of civilian inventors and engineers to advise the navy about the possible utilization of its own inventions and the inventions of other Americans submitted to the board. The Naval Consulting Board, initially headed by Edison, was formed in July 1915. It was from such simple beginnings and sweeping assumptions and assessments that the so-called military industrial complex emerged. Though the achievements of the Naval Consulting Board were limited for a number of reasons, the experience gained with it reinforced the proposition that the armed forces needed the help of civilian invention, research, and development in peace as well as in war.

What Is Expected of Naval Board

EDWARD MARSHALL

Secretary Daniels says great inventors will originate ideas and discuss those of others.

To give the nation's navy the benefit of the best thought of the nation's great civilian experts is the plan of Secretary Daniels, and his device for doing it is revolutionary.

By the frankly grateful statement of the Secretary himself, one of THE NEW YORK TIMES interviews, in which Thomas A. Edison gave his views of what America should do to make herself invincible, and, therefore, peaceful, was responsible for the project of establishing as an adjunct of the Navy an Advisory Board of great civilians.

Of this board Mr. Daniels named one member, Mr. Edison, asking the technical societies of the nation to supply the others by their own political non-partisan election. His hope was and is that thus may be accomplished an important detail of the interesting job of keeping the United States bright at a time when all the balance of the world is stained with battle-blood and dimmed by battle-smoke.

The week before it had been my great good fortune to have the epoch-making talk with Mr. Edison which resulted in the creation of this plan, I had discussed with Mr. Daniels his own hopes for the navy.

After the new plan was announced it seemd most likely that the fresh development might well have altered some of these, so I went to Washington again to talk things over once more with the Secretary in the light of new events.

"What do you expect of the new board?" I asked at once.

"Two things," said Mr. Daniels. "I expect it both to originate ideas and to examine those of others, critically and with

---

The New York Times Magazine, August 8, 1915, pp. 14–15.

such a concentration of ability as never has existed in the personnel of any other board.

"I shall not be disappointed. This board of great civilians will add more to the strength and efficiency of our navy than any one idea outside the service ever has added to the strength of any navy.

"Puzzling mechanical problems continually arise in these days of highly specialized fighting devices. Certainly we can hope more effectively to solve these problems when we have concentrated in the makeup of a Naval Advisory Board the nation's very greatest civilian experts in machines to work in harmony with our ablest naval men and for the self-same object.

"It is a curious fact that the world's civilian citizens often have felt themselves shut out from making contributions of ideas to their Governments.

"Sometimes they have made them, but it has been the general impression that every effort has been made to render this a difficult process. No one even can roughly guess at what the Governments have lost through lack of real co-operation between civilian experts and the navies of their nations.

"It is interesting to recall the fact, in this connection, that Ericsson was a civilian and that when he proposed his ironclad, the famous Monitor, which sank the Confederate Merrimac in Hampton Roads, there was much opposition to a Government trial of the idea.

"I do not think it in the least improbable that if our Government at the time had not been engaged in war and fairly desperate for fighting machines he might have found it utterly impossible to get a trial for his epochmaking invention.

"But the problems of this naval day are far more complicated than were the problems of that naval day. Sea fighting has developed into a highly specialized science, in which men must fail unless they be learned in the science of an exactly scientific profession. Bravery, of course, is as great a trait as ever, but in these days the merely brave man could not be the most effective. We must have upon our ships men who are not only brave but highly educated in their very technical work.

"The present European war has revealed a startingly complete change in naval methods.

"The great battle between the mightiest fighting ships the

world ever has known, which was confidently thought to be a certainty of the next war two years ago, has not occurred.

"It is most unlikely that it will occur.

"But the submarine, which, like the Monitor, was regarded as a possibility, but not as a certainty, has proved to be of great importance. It has slipped up to the side of warships and other craft of many times its size and sent them to the bottom, it has so demonstrated its effectiveness that great flotillas of the world's most imposing warships never have been sent out of their harbors, but lie in them, rusting, while the contest rages, guarded closely by torpedo nets against an undersea attack.

"Naval men have played their admirable part in the development of submarines both in this country and abroad, but, none the less, in its inception the submarine is a civilian and not a naval product.

"Holland and Lake, both of our own nation, neither of them ever in any way connected with the navy, have been the individuals to do most for its perfection. Holland died before he saw his dreams come wholly true, but Lake, today, is actually building submarines at his own shipyards.

"The Government today also is building for itself a submarine at its own shipyards, but in the main it is building it upon the lines which these two civilian geniuses developed as the result of their great concentration upon the single problem, unhampered by attention to the general details of the naval service. The Government in building is paying Mr. Lake for the use of his patented devices. But the Government has moved slowly and finds that this has handicapped it. For the Government-built submarine it will be unable to get, for love or money, the best-known engine within less than a year's time.

"Both the aeroplane and its development, the hydroaeroplane, are devices of civilian origin. While Government brains —and it must be thoroughly understood that nothing I may say about the need of co-operation of civilian brains with those of naval men is a reflection upon naval brains—have done much toward their perfection, civilians have done far more. That has been inevitable, and similar facts will be inevitable in future.

"These riders of the air and joint riders of the air and water need new engine ideas, too, yet under the system which has been pursued we have not found them. There are but a few concerns which manufacture engines, either for the underwater craft or aircraft, and their hands are overfull.

"They have done well for us, and we have done well for ourselves, but still we are confronted by unsolved engine problems of immense importance.

"Is it not obvious that we shall be the gainers when this civilian board co-operates with us?

"Certainly this great nation is capable of solving that annoying problem of submarine and air motive power; but it is of such importance that, as certainly, we must put all the nation's ablest engine-planning brains at work upon it. In that direction, to my mind, and not in the direction of a studied refusal to seek wisdom which may lie outside of a Government department must lie our national security.

"Another problem of immense importance, upon which might rest the fate of some momentous day if this nation unfortunately ever should be forced to go to war, is that of the torpedo.

"When one is devised which can be fired beneath the water and give no surface indication of its passage as it speeds grimly to its mark, the secrecy and, therefore, the effectiveness of the submarine will be immensely multiplied.

"For a long time the naval experts of the world have been working upon this. They have not solved the puzzle. The civilian board and the great interest in naval problems which it will arouse among civilian inventors may easily result in the discovery of some device which will accomplish that.

"Explosives offer another group of problems which must be carefully considered, not only by the admirable brains which, already, we have concentrated in our naval personnel, but by all the best brains in the nation.

"These few rough statements of some unsolved questions which most need solution will give a slight idea of the great importance of the work for national defense which now remains undone, not only here but in the balance of the world.

"The referendum feature of my plan will consist in placing these and other similar questions before the members of the board. I think the initiative, the announcement of newly created ideas, soon will follow in most instances.

"The eminent men who will make up the board will consider what is laid before them, and then, I think, we shall begin to see the working of the plan's initiative feature.

"I do not expect these celebrated and very busy men personally to devote much time to the questions of the board. But an immense advantage to the nation will accrue if we succeed in planting our most vital problems in these minds.

"We know that soil is fertile. We know that, for already it has produced mechanical and scientific marvels which have made us the world's wonder. Then it is not unreasonable, is it, to hope that such seed planted in such soil will grow and presently will fructify in great ideas which will make us so secure against attack that none will ever run the risk of trying to annoy us?

"The big thing about the whole idea of the board, I think, it is that it will bring into the definite service of our Government the men whose brains could not be purchased for such work by any conceivable expenditure of money.

"Their payment will lie in their realization of the highest patriotic service. When I went to Mr. Edison to ask him to accept the Chairmanship, I told him that I did not ask him for myself, that I did not even ask him for the navy, but that I asked him for the whole United States.

"That pleased him, I imagine, and it was absolutely true.

"Then, after he had told me that he would accept, I asked him if he thought that other eminent men who might be asked to work with us would be likely to accept the invitation.

" 'Why, yes, I'm sure they will,' said he. 'In this country no official honors have been provided for inventors or for other scientists. There are official honors waiting for the lawyer in the nation's Judgeships—and so on. But for the scientist there are none.

" 'This board supplies an honor of the highest order. Every man selected for its membership will feel as might an Englishman who had received a knighthood.'

"That was very gratifying.

"I am sure that men thus animated will accomplish wonders in creating helpful thought.

"Fulton thought about the steamship when no naval man had done so.

"When we began to build the Monitor both naval and civilian experts said the Government was wasting money. When it accomplished its great feat they criticised the Government because it had not built more monitors.

"Under the stimulation of the certainty of a sympathetic civilian body to consider them, other ideas as startling as the great thought of the turret ironclad surely will come to the surface, the inventions of great brains, today as utterly unknown as Ericsson's was then.

"Money could not do, I think, what thus this board will do. No one ever stimulated minds to great inventions by giving known inventors money and telling them to take it, go to some secluded spot and there invent new wonders. We must bring out the ideas which, unbidden, arrive in the creative mind—the inspirations.

"Most of the men who will comprise the board have spent their splendid energies at the invention of the instruments of peace, not warfare. That is as it should be.

"It is to be regretted that conditions ever should have risen which must request of them a change in the productive channels of their thought, diverting it toward warfare.

"But since this has occurred, and since military and naval conditions have swerved so far away from those which we have spent so lavishly of time and money in preparing for, since they have turned so generally from the old-time paths into the modern paths of chemistry and of electricity, we must have as our advisers the most thoughtful, most progressive, most developed men.

"Modern warfare is a very different affair from that of old. No bands with music thrill the hearts of Europe's fighting men on land or sea today.

"The glitter and the pomp of brilliant uniforms are gone forever.

"One chemist, one electrician might be greater in the warfare of the future than Napoleon, at his best, was in the warfare of the past.

"One scientist very probably may do more for the United States than any Admiral or General could do.

"It is interesting and very gratifying to find how quickly the thought of this consulting board has captured the people's thought throughout the country.

"The letters have poured in from every sort of people. The Vice President has sent me one. It said the plan caught his imagination, for he favored no armament save for our own defense, but that that must be complete.

"That is the American sentiment which most frequently has been expressed, from every section of the nation.

"I have a barrelful of letters and editorial clippings. I am saving these to read while I am on vacation and have not the slightest doubt that in them will be found many suggestions of great value, for it is plain enough that, today, owing to this plan, thousands of brains are working at our national problem, which, previous to the publication of your interview with Mr. Edison, had not given any thought to it.

"I have no doubt that I shall find among these letters many from the nation's so-called 'cranks.' But what is a crank? The crank of today is the genius of tomorrow. Our 'cranks' have done great things. They have done harmful things, sometimes, but they have done some very useful things. The few letters which will prove to be unwise won't bother me at all.

"We have gone through enough of these communications to make us feel certain of the average value. Some of them will be among the first things to be brought to the attention of the new Advisory Board.

"One, from Colorado, suggests an improvement in wireless telegraphy. We at once discussed it with the head of our department, who was very much impressed by it.

"We then asked the man to come to Washington, but it seems that he has not the funds, and we have none from which at present we can draw.

"You see? That man, without funds for the journey to Washington and without funds for experimentation, may have a very great idea. That is a situation which doubtless is recurring constantly. It is an evil and a waste, and the existence of the board will do much, I hope, to do away with its repetition in the future.

"It is evident that we need a national laboratory. Its value to the nation easily might be inestimable.

"Some years ago we established at Annapolis a laboratory for coal testing. In one year these coal tests saved the Government as much as it will cost to run the laboratory for many years.

"We now take for the navy only coal from mines which have been tested and which we know produce coal sure to burn effectively and economically in naval ships. Through this laboratory we have discovered that the coal of Government-owned mines up in Alaska is good for naval vessels, and through its tests we have discovered many valuable things concerning oil as fuel.

"But it is comparatively small and narrow in its scope. What we really need is a great institution for close chemical analysis which will cost from half a million to a million dollars, and which will be supported by an annual appropriation to be expended in accordance with the directions of the new Advisory Board in conjunction with the Secretary of the Navy.

"The board, too, should be allowed to compensate originators of good ideas, even though these men be in the navy. I think this is imperative and mere justice. We have men in the navy who have done extraordinary things, but it is conceivable that they might do things more wonderful if they could be paid for what they do and relieved of other work, for that procedure would enable them to concentrate on the production of more big ideas and upon the full development of those which they already have produced.

"American genius will meet any needs and meet them as they have not been met by the conventional methods of the older nations. Always we have been creative people and original and daring people. But, perhaps, others have had more value from our own originality than we ourselves have had. That is certainly a failing of our system.

"Think of the submarine—purely an American idea! We have done less with it than any other of the really great nations. Think of the folly of it!

"Suppose that, through governmental aid, we had brought to its perfection the aeroplane, which is absolutely an American idea. How vast would be our present-day advantage.

"But what really occurred?

"When Langley got some public money with which to build the first one, he could not get enough, and so it failed and was discredited. He died a broken-hearted man.

"Yet we now know that Langley was a genius of high order, and a practical genius, too. His ideas were exactly right, so far as flying was concerned.

"His device lacked only motive power. When Curtiss recently put a modern engine into the old Langley flying machine

—the one whose failure crushed his soul and heaped the ridicule of fools on him—the apparatus flew as Langley had declared it would fly.

"Such a board as now is under way toward an effective formation, as the result of your remarkable interview with Mr. Edison, will tend to mobilize the genius of the country and will pay for its slight cost a thousand times a year in public service after once it is in operation."

PART IV

Ambivalence, Technocracy, and Scientism

INTRODUCTORY NOTE

The carnage of World War I and the well-publicized exploitation of technology brought a new ambiguity, the legacy of a century of rarely qualified praise confronted by the obvious horrors of total war. It was further reinforced by the growing belief that technology was not directed by benign providence or inexorable laws of progress toward social objectives believed generally—and vaguely—desirable. The growing conviction that technology and progress were not synonyms and that man needed to assume control of, and make decisions about, the use of technology brought less euphoric prophesies for the future.

This ambiguity is explicit in "Science to End War or End the Race," the unsigned *Literary Digest* article that follows. After predicting the effects of unlimited chemical warfare, of unrestrained aerial bombardment, and of unprecedentedly accurate artillery and rifle fire, the author ventured to hope that fear might be the beginning of wisdom. The fear of terrible weapons, he hoped, would bring disarmament and peace.

The essayist believed that scientists and engineers were publicizing the effects of the weapons used in the war in the hope of making the very thought of war intolerable. He reported that well-known inventors and scientists—like the American Leo Baekeland, the chemist whose most widely publicized invention was Bakelite—evangelistically declared that chemists "ought to turn to and help steer the world away from a frightful chemical debacle." The author did not go so far as to suggest that the scientists were intentionally "making war an engine of world destruction" so that this threat might force the world to disarm or perish in ruins, but he clearly anticipated that this might well be the result. Three decades later scientists would again publicize the destructiveness of new weapons—in this case, atomic—in the hope that war would become unthinkable.

Science to End War or End the Race

An argument for peace which was badly shaken by the advent of the World War is being advanced again with renewed confidence as a result of hints dropt recently by scientists and technicians. This argument is that the appalling and ever-increasing destructiveness of the weapons of modern warfare tends to make the very thought of war intolerable. The human race must choose between peace and suicide, say the advocates of this theory, for unless some way is found to prevent wars, war will destroy civilization itself. This warning is given new weight by statements made before two great gatherings of scientists—a convention of a thousand chemists in Ithaca, New York, and the centenary celebration of the Franklin Institute in Philadelphia. At the Ithaca meeting Sir Max Muspratt, one of the most eminent chemists of Great Britain, declared that the science of chemistry has now reached the stage where it is able to destroy the world in short order; and Dr. Leo Hendrik Baekeland, president of the American Chemical Society, averred that the complete destruction of entire cities was merely a matter of somebody giving the order with such authority that the chemists would have to obey. "And then," notes the Philadelphia *Evening Public Ledger*, "both Sir Max and Dr. Baekland grew evangelistic and declared that the chemists ought to turn to and help steer the world away from a frightful chemical debacle."

At the international conference of scientists in Philadelphia more specific testimony was offered concerning the increasing horrors of modern warfare. General Squier declared that no limit could be imposed upon new ideas of attack if war should come again; and he mentioned as a possibility that "a whole nation will be put to sleep for twenty-four hours by gases distributed by radio-controlled airplanes." And Gen. Mason M. Patrick, Chief of the Army Air Service, who conceived and

---

The Literary Digest, 183 (October 24, 1924), 13.

fostered the round-the-world flight, gave the following testimony concerning the progress made in fighting from the air:

> "Airplanes by dropping bombs can sink any naval vessel yet built. Even when a direct hit is not made, a bomb dropt alongside gives a vital blow. Anti-aircraft guns are powerless to protect.
>
> "Coast-defense scouting planes could discover the approach of an enemy fleet 200 miles off the coast. That would mean eight to ten hours to concentrate an air force.
>
> "With the aid of the Chemical Service we have developed a means of laying smoke-screens to prevent ships from seeing planes.
>
> "Ships are powerless against night attacks from airplanes. The planes can drop parachute flares which illuminate the sea for miles. Above the source of light the planes remain invisible. Planes now can talk to one another from ten miles apart or talk back and forth with a land station 100 miles away by radio.
>
> "We have developed automatic pilots so that airplanes can be sent off at a predetermined height and made to drop their bombs at a predetermined point. Or they can be maneuvered by radio with no pilot aboard. They may be carried across a continent by dirigibles and then sent forth on combat missions."

The same body of scientists were informed by Maj-Gen. C. O. Williams, Chief of Army Ordnance, that the range of our guns has been virtually doubled since the World War. His statements, which are reported to have "made his hearers gasp," are briefly summarized as follows by a Philadelphia correspondent of the New York *Herald Tribune:*

> "A semi-automatic rifle is in process of perfection for use as a shoulder weapon, General Williams said, as well as a .50 caliber machine-gun to replace the war-time .30 caliber. New trench mortars of great firing accuracy; high-speed tanks, with guns in turrets, instead of in the body of the tank; a 75-millimeter field-gun of 15,000 yards range, compared to 9,000 in the war, and other wonders were outlined.
>
> "The 105-millimeter gun has been replaced with a 155-millimeter, General Williams said, increasing the range from 12,000 to 20,000 yards. The 4.7-inch gun now hurls a projectile of 100 pounds 20,000 yards, whereas a 45-lb. shell was formerly used, he continued. . . .
>
> "Modern aerial bombs, he stated, are six times more

> destructive than those used in the Zeppelin raids of the World War."

Such suggestions as the bombing of cities by manless airplanes, the destruction of a battle-ship by a single air bomb, and the paralyzing of an entire nation by a combination of chemicals, remarks the New York *Times*, give "an intimation of what may come upon this earth if the Assembly sitting in Geneva or some other international effort does not succeed in averting another world war." And in the New York *Evening Post* we read:

> "Scientists are doing more for the cause of peace and a warless world than all our statesmen, leagues and peace societies. They made disarmament the great world problem by making war an engine of world destruction. This threat raised by Science has become so great that it may be on the eve of forcing the world to solve it or perish in its own ruins. . . .
>
> "Applied Science has brought aviation to a day when a lone aviator, riding a mother plane, may drive before him through the sky-lanes a covey of death in the shape of pilotless planes. Under gas waves spread by them a great city or an Army might be anesthetized for a day or sent into that sleep that knows no waking. In this winged brood of destruction, radio-guided, every plane will be able to drop bombs twenty times as destructive as the largest shell ever hurled from a gun muzzle. . . .
>
> "Science has left no non-combatants in modern warfare, which has become a clash of nations rather than of Armies. It has lifted war from the land and the water, and from under the water, into a 'Fourth Dimension,' the air. Ancient conquerors ravaged the land with the sword and the torch and sowed with salt the ruined towns of their enemies. A modern conqueror hurls tons of nitrogen explosives at a nation and sows an invisible death out of the sky.
>
> "The wars where Science is the handmaiden of Mars put a new fear into mankind and this fear may be the beginning of wisdom. Certainly the world is more afraid of the 'next war' than ever before. Mankind begins to dread that Science may solve all the problems of war and disarmament by the final and utter destruction of the race and its civilization. In the shadow of that fear the instinct of self-preservation in humanity begins to assert itself as it glimpses the new terrors that are the products and by-products of Science."

INTRODUCTORY NOTE

The British Bishop of Ripon touched an American nerve made sensitive by the use of technology in the Armageddon of 1914–1918. In "Is Scientific Advance Impeding Human Welfare?" his considerable doubts about scientific advance contributing to human welfare and his call for a ten-year holiday in science are evaluated. Although his remarks were made from a Leeds pulpit, they caused sympathic reverberations in the popular American press. The *Brooklyn Eagle, The New York Times,* the *New York Sun,* and the *Providence Journal* all took the remarks as deserving serious comment. A century earlier the bishop would probably have been ignored or ridiculed on this side of the Atlantic, but in 1927 the *Chicago Evening Post,* in a mood reflective of a growing ambiguity toward technology and applied science, felt that the Bishop's views were shared by multitudes of Americans.

A rejoinder to the bishop by the American physicist Robert Millikan pointed out the absolute necessity of scientific research and technological development in the medical sciences and technology: "The suggestion of a ten-year holiday from research work for the lessening of human suffering is ridiculous." If this attitude —probably shared by most Americans—was linked to Millikan's statement that medical science was dependent on the sciences of chemistry and physics, the same sciences that had contributed to the development of deadly weapons like poison gas and remote-controlled bombs, the American reader must have been made aware of the dilemma confronting scientist and layman alike.

Is Scientific Advance Impeding Human Welfare?

Admiration, mixed with some amazement at his temerity, greets the Bishop of Ripon's suggestion that science take a ten-year holiday to give us time to assimilate the knowledge we already have. The Bishop was speaking at a church service in Leeds, to which had come many men of distinguished name who were in the city to attend a meeting of the British Association for the Advancement of Science. Many new discoveries and important steps in the advancement of scientific knowledge—sufficient to daze the ordinary mind—had been announced, and the Bishop thinks that so much is being done toward material progress that man is in danger of losing his own soul. Himself a man of deep learning, his suggestion arouses much comment no less on this side of the Atlantic than on the other. "The idea is an old one," observes the Brooklyn *Eagle*, but the remedy the Bishop proposes is impossible of application. "Laws will not prevent men from thinking. Even a resolution of the British Association would not prevent scientists here and there from following their investigations out in secret—bootleg investigations, so to speak."

Nevertheless, the Bishop thinks, as his words are quoted in a dispatch to the New York *Times*, that "with all his new mastery over nature, man has not seemed really to be advancing his own cause. The development of his resources has not helped either development or happiness for himself. Until this disproportion is somehow rectified, man can not feel safe, and the very greatness of his recent achievements would seem to make his ruin more certain and more complete." As the Bishop sees it:

> "We could get on very much more happily if aviation, wireless, television and the like were advanced no further than at present.
>
> "Dare I even suggest, at the risk of being lynched by some

The Literary Digest, 95 (October 1, 1927), 13.

> of my hearers, that the sum of human happiness, outside of scientific circles, would not necessarily be reduced if for, say, ten years every physical and chemical laboratory were closed and the patient and resourceful energy displayed in them transferred to recovering the lost art of getting together and finding a formula for making the ends meet in the scale of human life?
>
> "It would give 99 per cent. of us who are non-scientific some chance of assimilating the revolutionary knowledge which in the first quarter of this century 1 per cent. of the explorers have acquired. The 1 per cent. would have leisure to read up on one another's work; and all of us might go meanwhile in tardy quest of that wisdom which is other than and greater than knowledge, and without which knowledge may be a curse.
>
> As things stand to-day, we could get on without further additions for the present to our knowledge of nature. We can not get on without a change of mind in man."

The first retort to the Bishop came from Sir Oliver Lodge, who, addressing another meeting, said that the suggestion that the world was going too fast reminded him of the view-point taken by his grandmother, who protested at trains traveling forty miles an hour. "It is not possible to call a halt," Sir Oliver declared, we read in the *Times* dispatch. "If we stopt, the world would go to pieces." Referring to the fact that many people dislike the idea that they are related to the lower animals, Sir Oliver said he regarded such evolutionary advance as inspiring. Had man descended from the angels, he said, it would have been depressing. Nothing can stand still, replied Sir Arthur Keith, whose presidential address before the scientists supporting Darwin's theory of descent has attracted wide notice. The Church itself, said Sir Arthur, would perish if it remained stagnant.

To come to this side of the water, the New York *Sun* thinks that the Bishop "started something and will not soon hear the end of it." Theoretically, observes the Manchester *Union*, the churchman's statement has a certain plausibility, since the gains of science during the past twenty-five years are "truly overwhelming to the lay mind. . . . Science to-day has given us a new world, requiring a new system of knowledge for its interpretation, and one may well suggest the advisability of a holiday for orientation and adjustment."

"But viewed from another angle this statement becomes absurd. For while science has a theoretical bearing, it has also a practical bearing, and from the latter standpoint the proposal to close our scientific laboratories for ten years is unthinkable. Science to-day has found its way into every field of human activity. Industry, agriculture, transportation, and philanthropy are all controlled and dominated by it. A glance at one branch of human operation will make this plain, that is the field of medicine. During the past twenty-five years the medical laboratory has discovered the secret of a multitude of diseases, and as a result the dread and power of these diseases have been broken. The suggestion of a ten-year holiday from research work for the lessening of human suffering is ridiculous. This is only one case in many that might be cited when the practical bearings of science are taken into account.

"It is of course that an excessive zeal for scientific research always reacts against interest in philosophy. In this respect our age is not the exception. We have strest the practical bearings of science somewhat at the expense of the theoretical. We must frankly admit that our philosophy of life finds itself staggered by present-day scientific discoveries, and only with hesitation and often with protest, is ready to accept them. But this is no reason for calling a halt to science, but rather a demand for the advancement of philosophy. It is not probable that a stoppage in scientific research would greatly promote those intellectual pursuits that make for a more thorough and up-to-date interpretation of life.

"Just now there is an apparent conflict between science and philosophy. The old interpretation of life is hesitant to abandon cherished notions, even tho they are out of harmony with new truth. But this conflict is in interpretation only, and it can readily be removed by an earnest attempt to coordinate facts as they are in the search for truth."

Pondering the question whether religious leaders have shown a disposition to get together with the scientists in understanding and expounding the theory of evolution, the Providence *Journal* declares that they still look upon the sermons on evolution preached forty years ago by Henry Ward Beecher "as the voice of one crying in the wilderness," and says, "of course there will be no moratorium or 'scientific holiday.' The world can not afford it."

Nevertheless, asserts the Chicago *Evening Post,* the feeling which impels the Bishop of Ripon to make his suggestion is shared by multitudes of men, since "science has been leading us rather a giddy chase for the last two or three decades." Not only are laymen far behind in scientific knowledge, we are told; but even the scientists themselves are, in certain departments, far behind in the preliminary work of classifying new facts. "And the danger to the world lies in the fact that it is largely these muddled minds, glib or giddy, which are making the rules—called laws and treaties—by which civilization is trying to regulate itself."

INTRODUCTORY NOTE

The pessimism of Bertrand Russell, the British philosopher, mathematician, and social critic, expressed in the following essay differs dramatically from earlier American optimism. After World War I Americans were becoming more critical of science and technology, but, at least for a time, Russell seems to have articulated this reaction more effectively than Americans. His ideas were widely printed in American journals in the mid-twenties, and he toured America to give lectures. Appearing as he did in American periodicals and before American audiences, it seems reasonable to consider him as articulating and formulating American attitudes toward science and technology.

Russell's attitudes were shaped by the shock of seeing technology and science used in total war. Englishmen had known the destructiveness and had suffered the casualties more intimately and more fully than Americans. But the war, though more remote for Americans, prepared them intellectually for ideas such as Russell's that "science threatens to cause the destruction of our civilization."

Russell's sustained argument in "The Effect of Science on Social Institutions" anticipated many of the attitudes encountered almost a half-century later. His concern that science and technology would stimulate more intense international economic competition has become a commonplace; his fear that science and technology, after having been used to sustain increased population and raise the living standard, would be used in war to decimate the population and lay waste to the land has also become a commonplace. Russell's prophesy that communication and transportation technology would bring ever more complex and large bureaucracies, military and civilian, has come to pass; his statement that centrally directed military and industrial bureaucracies could submerge individual freedom was uncomfortably close to the mark. Finally, Russell's insistence that powerful technology had fallen, and would fall, into the hands of irrational, conscienceless, power-infatuated men was a theme shared by social critics writing years later.

Bertrand Russell (1872–1972), an English philosopher who had a reputation as a mathematician, was also active as a social critic, especially in the cause of peace. He taught at Cambridge, Harvard, Peking, and other universities. Among his books are *Principia Mathematica* (with A. N. Whitehead), 3 vols. (1910–1913); *The Prospects of Industrial Civilisation* (1923); *Icarus, or The Future of Science* (1924); and *A History of Western Philosophy* (1945).

The Effect of Science on Social Institutions

BERTRAND RUSSELL

The world in which we live differs profoundly from that of Queen Anne's time, and this difference is mainly attributable to science. That is to say, the difference would be very much less than it is but for various scientific discoveries, and but for what has resulted from those discoveries by the operation of ordinary human nature. The changes that have been brought about have been partly good, partly bad; whether, in the end, science will prove to have been a blessing or a curse to mankind is, to my mind, still a doubtful question.

A science may affect human life in two different ways. On the one hand, without altering men's passions or their general outlook, it may increase their power of gratifying their desires. On the other hand, it may operate through an effect upon the imaginative conception of the world, the theology or philosophy which is accepted in practice by energetic men. The latter is a fascinating study, but I shall almost wholly ignore it, in order to bring my subject within a manageable compass. I shall confine myself largely to the effect of science in enabling us to gratify our passions more freely, which has hitherto been far the more important of the two.

From our point of view, we may divide the sciences into three groups: physical, biological, and anthropological. In the physical group I include chemistry, and broadly speaking, any science concerned with the properties of matter apart from life. In the anthropological group I include all studies specially concerned with man: human physiology and psychology (between which no sharp line can be drawn), anthropology, history, sociology and economics. All these studies can be illuminated by considerations drawn from biology; for instance, Rivers threw a new light on parts of economics by adducing facts about landed property among birds during the breeding

---

The Survey, Graphic Number 52, (April 1, 1924), 5–11.

season. But in spite of their connection with biology—a connection which is likely to grow closer as time goes on—they are broadly distinguished from biology by their methods and data, and deserve to be grouped apart, at any rate in a sociological inquiry.

The effect of the biological sciences, so far, has been very small. No doubt Darwinism and the idea of evolution affected men's imaginative outlook; arguments were derived in favour of free competition and also of nationalism. But these effects were of the sort that I propose not to consider. It is probable that great effects will come from these sciences sooner or later. Mendelism might have revolutionized agriculture, and no doubt some similar theory will do so sooner or later. Bacteriology may enable us to exterminate our enemies by disease. The study of heredity may in time make eugenics an exact science, and perhaps we shall in a later age be able to determine at will the sex of our children. This would probably lead to an excess of males, involving a complete change in family institutions. But these speculations belong to the future.

The anthropological sciences are those from which, a priori, we might have expected the greatest social effects, but hitherto this has not proved to be the case, partly because these sciences are mostly still at an early stage of development. Even economics has not so far had much effect. Where it has seemed to have, this is because it advocated what was independently desired. Hitherto, the most effective of the anthropological sciences has been medicine, through its influence on sanitation and public health, and through the fact that it has discovered how to deal with malaria and yellow fever. Birth-control is also a very important social fact which comes into this category. But although the future effect of the anthropological sciences (to which I shall return presently) is illimitable, the effect up to the present has been confined within fairly narrow limits.

One general observation to begin with. Science has increased man's control over nature, and might therefore be supposed likely to increase his happiness and well-being. This would be the case if men were rational, but in fact they are bundles of passions and instincts. An animal species in a stable environment, if it does not die out, acquires an equilibrium between its passions and the conditions of its life. If the conditions are suddenly altered, the equilibrium is upset. Wolves in a state of nature have difficulty in getting food, and therefore need the stimulus of a very insistent hunger. The result is that their descendants, domestic dogs, over-eat if they are allowed to do so. When a certain amount of something is useful, and the difficulty of obtaining it is diminished, instinct will usually lead an animal to excess in the new circumstances. The sudden change produced by science has upset the balance between our instincts and our circumstances, but in directions not sufficiently noticed. Over-eating is not a serious danger, but over-fighting is. The human instincts of power and rivalry, like the dog's wolfish appetite, will need to be artificially curbed if industrialism is to succeed.

Much the greatest part of the changes which science has made in social life is due to the physical sciences, as is evident when we consider that they brought about the industrial revolution. This is a trite topic, about which I shall say as little as my subject permits. There are, however, some points which must be made.

First, industrialism still has great parts of the earth's surface to conquer. Russia and India are very imperfectly industrialized; China hardly at all. In South America there is room for immense development. One of the effects of industrialism is to make the world an economic unit; its ultimate consequences will be very largely due to this fact. But before the world can be effectively organized as a unit, it will probably be necessary to develop industrially all the regions capable of development that are at present backward. The effects of industrialism change as it becomes more wide-spread; this must be remembered in any attempt to argue from its past to its future.

The second point about industrialism is that it increases the productivity of labor, and thus makes more luxuries possible. At first, in England, the chief luxury achieved was a larger population, with an actual lowering of the standard of life. Then came a golden age when wages increased, hours of labor diminished, and simultaneously the middle-class grew more prosperous. That was while Great Britain was still supreme. With the growth of foreign industrialism, a new epoch began. Industrial organizations have seldom succeeded in becoming world-wide, and have consequently become national. Competition, formerly between individual firms, is now mainly between nations, and is therefore conducted by methods quite different from those contemplated by the classical economists.

Modern industrialism is a struggle between nations for two things, markets and raw materials, as well as for the sheer pleasure of dominion. The labor which is set free from providing the necessaries of life tends to be more and more absorbed by national rivalry. There are first the armed forces of the state; then those who provide munitions of war, from the raw minerals up to the finished product; then the diplomatic and consular services; then the teachers of patriotism in schools; then the press. All of these perform other functions as well, but the chief purpose is to minister to international competition. As another class whose labors are devoted to the same end, we must add a considerable proportion of the men of science. These men invent continually more elaborate methods of attack and defence. The net results of their labors is to diminish the proportion of the population that can be put into the fighting line, since more are required for munitions. This might seem a boon, but in fact war is now-a-days primarily against the civilian population, and in a defeated country they are liable to suffer just as much as the soldiers.

It is science above all that has determined the importance of raw materials in international competition. Coal and iron and oil, especially, are the bases of power, and thence of wealth. The nation which possesses them, and has the industrial skill required to utilize them in war, can acquire markets by armed force, and levy tribute upon less fortunate nations. Economists have underestimated the part played by military prowess in the acquisition of wealth. The landed aristocracies of Europe were, in origin, warlike invaders. Their defeat by the bourgeoisie in the French Revolution, and the fear which this generated in the Duke of Wellington, facilitated the rise of the middle class. The wars of the eighteenth century decided that England was to be richer than France. The traditional economist's rules for the distribution of wealth hold only when men's actions are governed by law, i. e., when most people think the issue unimportant. The issues that people have considered vital have been decided by civil wars or wars between nations. And for the present, owing to science, the art of war consists in possessing coal, iron, oil, and the industrial skill to work them. For the sake of simplicity, I omit other raw materials, since they do not affect the essence of our problem.

We may say, therefore, speaking very generally, that men have used the increased productivity which they owe to science for three chief purposes in succession: first, to increase the population; then, to raise the standard of comfort; and finally, to devote more energy to war. This last result has been chiefly brought about by competition for markets, which led to competition for raw materials, especially the raw materials of munitions.

The stimulation of nationalism which has taken place in modern times is, however, due very largely to another factor, namely the increase of organization, which is of the very essence of industrialism. Wherever expensive fixed capital is required, organization on a large scale is of course necessary. In view of the economies of large-scale production, organization in marketing also becomes of great importance. For some purposes, if not for all, many industries come to be organized nationally, so as to be in effect one business in each nation.

Science has not only brought about the need of large organizations, but also the technical possibility of their existence. Without railways, telegraphs, and telephones, control from a centre is very difficult. In ancient empires, and in China down to modern times, provinces were governed by practically independent satraps or proconsuls, who were appointed by the central government, but decided almost all questions on their own initiative. If they displeased the sovereign, they could only be controlled by civil war, of which the issue was doubtful. Until the invention of the telegraph, ambassadors had a great measure of independence, since it was often necessary to act without waiting for orders from home. What applied in politics applied also in business; an organization controlled from the centre had to be very loosely knit, and to allow much autonomy to subordinates. Opinion as well as action was difficult to mould from a centre, and local variations marred the uniformity of party creeds.

Now-a-days all this is changed. Telegraph, telephone, and wireless make it easy to transmit orders from a centre; railways and steamers make it easy to transport troops in case the orders are disobeyed. Modern methods of printing and advertising make it enormously cheaper to produce and distribute one newspaper with a large circulation than many with small circulations; consequently, in so far as the press controls opinion, there is uniformity, and, in particular, there is uniformity of news. Elementary education, except in so far as religious denominations introduce variety, is conducted on a uniform pattern decided by the state, by means of teachers whom the state has trained, as far as possible, to imitate the regularity and mutual similarity of machines produced to standard. Thus the material and psychological conditions for a great intensity of organization have increased *pari passu,* but the basis of the whole development is scientific invention in the purely physical realm. Increased productivity has played its part, by making it possible to set apart more labour for propaganda, under which head are to be included advertisement, the cinema, the press, education, politics, and religion. Broadcasting is a new method likely to acquire great potency as soon as people are satisfied that it is *not* a method of propaganda.

Political controversies, as Graham Wallas has pointed out, ought to be conducted in quantitative terms. If sociology were one of the sciences that had affected social institutions (which it is not), this would be the case. The dispute between anarchism and bureaucracy at present tends to take the form of one side maintaining that we want no organization, while the other maintains that we want as much as possible. A person imbued with the scientific spirit would hardly even examine these extreme positions. Some people think that we keep our rooms too hot for health, others that we keep them too cold. If this were a political question, one party would maintain that the best temperature is the absolute zero, the other that it is the melting point of iron. Those who maintained any intermediate position would be abused as timorous time-servers, concealed agents of the other side, men who ruined the enthusiasm of a sacred cause by tepid appeals to mere reason. Any man who had the courage to say that our rooms ought to be neither too hot nor too cold would be abused by both parties, and probably shot in No Man's Land. Possibly some day politics may become more rational, but so far there is not the faintest indication of a change in this direction.

To a rational mind, the question is not: Do we want organization or do we not? The question is: How much organization do we want, and where and when and of what kind? In spite of a temperamental leaning to anarchism, I am persuaded that an industrial world cannot maintain itself against internal disruptive forces without a great deal more organization than we have at present. It is not the amount of organization, but its kind and its purposes, that cause our troubles. But before tackling this question, let us pause for a moment to ask ourselves what is the measure of the intensity of organization in a given community.

A man's acts are partly determined by spontaneous impulse, partly by the conscious or unconscious effects of the various groups to which he belongs. A man who works (say) on a railway or in a mine is, in his working hours, almost entirely determined in his actions by those who direct the collective labor of which he forms part. If he decides to strike, his action is again not individual, but determined by his union. When he votes, party caucuses have limited his choice to one of two or three men, and party propaganda has induced him to accept *in toto* one of the two or three blocks of opinions which form the rival party programmes. His choice between the parties may be individual, but it may also be determined by the action of some group, such as a trade union, which collectively supports one party. His newspaper-reading exposes him to great organized forces; so does the cinema, if he goes to it. His choice of a wife is probably spontaneous, except that he must choose a woman of his own class. But in the education of his children he is almost entirely powerless; they must have the education which is provided. Organization thus determines many vital things in his life. Compare him with a handicraftsman or peasant proprietor who cannot read and does not have his children educated, and it becomes clear what is meant by saying that industrialism has increased the intensity of organization. To define this term, we must, I think, exclude the unconscious effects of groups, except as causes facilitating the conscious effects. We may define the intensity of organization to which a given individual is subject as the proportion of his acts which is determined by the orders or advice of some group, expressed through democratic decisions or executive officers. The intensity of organization in a community may then be defined as the average intensity for its several members.

The intensity of organization is increased not only when a man belongs to more organizations, but also when the organizations to which he already belongs play a larger part in his life, as, for example, the state plays a larger part in war than in peace.

Another matter which needs to be treated quantitatively is the degree of democracy, oligarchy, or monarchy in an organization. No organization belongs completely to any one of the three types. There must be executive officers, who will often in practice be able to decide policy, even if in theory they cannot do so. And even if their power depends upon persuasion, they may so completely control the relevant publicity that they can always rely upon a majority. The directors of a railway company, for instance, are to all intents and purposes uncontrolled by the shareholders, who have no adequate means of organizing an opposition if they should wish to do so. In America, a railroad president is almost a monarch. In party politics, the power of leaders, although it depends upon persuasion, continually increases as printed propaganda becomes more important. For these reasons, even where formal democracy increases, the real degree of democratic control tends to diminish, except on a few questions which rouse strong popular passions.

The result of these causes is that, in conquence of scientific inventions which facilitate centralization and propaganda, groups become more organized, more disciplined, more group-conscious, and more docile to leaders. The effect of leaders on followers is increased, and the control of events by a few prominent personalities becomes more marked.

In all this there would be nothing very tragic, but for the fact, with which science has nothing to do, that organization is almost wholly national. If men were actuated by the love of gain, as the older economists supposed, this would not be the case; the same causes which have led to national trusts would have led to international trusts. This has happened in a few instances, but not on a sufficiently wide scale to affect politics or economics very vitally. Rivalry is, with most well-to-do energetic people, a stronger motive than love of money. Successful rivalry requires organization of rival forces; the tendency is for a business such as oil, for example, to organize itself into two rival groups, between them covering the world. They might, of course, combine, and they would no doubt increase their wealth if they did so. But combination would take the zest out of life. The object of a football team, one might say, is to kick goals. If two rival teams combined, and kicked the ball alternately over the two goals, many more goals would be scored. Nevertheless no one suggests that this should be done, the object of a football team being not to kick goals but to win. So the object of a big business is not to make money, but to win in the contest with some other business. If there were no other business to be defeated, the whole thing would become uninteresting. This rivalry has attached itself to nationalism, and enlisted the support of the ordinary citizens of the countries concerned; they seldom know what it is that they are supporting, but, like the spectators at a football match, they grow enthusiastic for their own side. The harm that is being done by science and industrialism is almost wholly due to the fact that, while they have proved strong enough to produce a *national* organization of economic forces, they have not proved strong enough to produce an international organization. It is clear that political internationalism such as the League of Nations was supposed to inaugurate, will never be successful until we have economic internationalism, which would require, as a minimum, an agreement between various national organizations dividing among them the raw materials and markets of the world. This, however, can hardly be brought about while big business is controlled by men who are so rich as to have grown indifferent to money, and to be willing to risk enormous losses for the pleasure of rivalry.

The increase of organization in the modern world has made the ideals of liberalism wholly inapplicable. Liberalism, from Montesquieu to President Wilson, was based upon the assumption of a number of more or less equal individuals or groups, with no differences so vital that they were willing to die sooner than compromise. It was supposed that there was to be free competition between individuals and between ideas. Experience has shown, however, that the existing economic system is incompatible with all forms of free competition except between states by means of armaments. I should wish, for my part, to preserve free competition between ideas, though not between individuals and groups, but this is only possible by means of what an old-fashioned liberal would regard as interferences with personal liberty. So long as the sources of economic power remain in private hands, there will be no liberty except for the few who control those sources.

Such liberal ideals as free trade, free press, unbiased education, either already belong to the past or soon will do so. One of the triumphs of early liberalism in England was the establishment of parliamentary control over the army; this was the *casus belli* in the Civil War, and was decided by the Revolution of 1688. It was effective so long as Parliament represented the same class from which army officers were drawn. This may cease to be the case before long. Russia, Hungary, Italy, Spain, and Bavaria have shown in recent years how frail democracy has become; east of the Rhine it lingers only in outlying regions. Constitutional control over armaments must, therefore, be regarded as another liberal principle which is rapidly becoming obsolete.

It would seem probable that, in the next fifty years or so, we shall see a still further increase in the power of governments, and a tendency for governments to be such as are desired by the men who control armaments and raw materials. The forms of democracy may survive in western countries, since those who possess military and economic power can control education and the press, and therefore can usually secure a subservient democracy. Rival economic groups will presumably remain associated with rival nations, and will foster nationalism in order to recruit their football teams.

There is, however, a hopeful element in the problem. The planet is of finite size, but the most efficient size for an organization is continually increased by new scientific inventions. The world becomes more and more of an economic unity. Before very long the technical conditions will exist for organizing the whole world as one producing and consuming unit. If, when that time comes, two rival groups contend for mastery, the victor may be able to introduce that single world-wide organization that is needed to prevent the mutual extermination of civilized nations. The world which would result would be, at first, very different from the dreams of either liberals or socialists; but it might grow less different with the lapse of time. There would be at first economic and political tyranny of the victors, a dread of renewed upheavals, and therefore a drastic suppression of liberty. But if the first half-dozen revolts were successfully repressed, the vanquished would give up hope, and accept the subordinate place assigned to them by the victors in the great world-trust. As soon as the holders of power felt secure, they would grow less tyrannical and less energetic. The motive of rivalry being removed, they would not work so hard as they do now, and would soon cease to exact such hard work from their subordinates. Life at first might be unpleasant, but it would at least be possible, which would be enough to recommend the system after a long period of warfare. Given a stable world-organization, economic and political, even if, at first, it rested upon nothing but armed force, the evils which now threaten civilization would gradually diminish, and a more thorough democracy than that which now exists might become possible. I believe that, owing to men's folly, a world-government will only be established by force, and will therefore be at first cruel and despotic. But I believe that it is necessary for the preservation of a scientific civilization, and that, if once realized, it will gradually give rise to the other conditions of a tolerable existence.

It remains to say something about the future effects of the anthropological sciences. This is of course extremely conjectural, because we do not know what discoveries will be made. The effect is likely to be far greater than we can now imagine, because these sciences are still in their infancy. I will, however, take a few points on which to hang conjectures. I do not wish to be supposed to be making prophecies; I am only suggesting possibilities which it may be instructive to consider.

Birth-control is a matter of great importance, particularly in relation to the possibility of a world-government, which could hardly be stable if some nations increased their population much more rapidly than others. At present birth-control is increasing in all civilized countries, though in most it is opposed by governments. This opposition is due partly to mere superstition or desire to conciliate the Catholic vote, partly to the desire for large armies and severe competition between wage-earners, so as to keep down wages. In spite of the opposition of governments, it seems probable that birth-control will lead to a stationary population in most white nations within the next fifty years. There can be no security that it will stop with a stationary population; it may go on to the point where the population diminishes.

This situation will lead to a tendency—already shown by the French—to employ more prolific races as mercenaries. Governments will oppose the teaching of birth-control among Africans, for fear of losing recruits. The result will be an immense numerical inferiority of the white races, leading probably to their extermination in a mutiny of mercenaries. If, however, a world-government is established, it may see the desirability of making subject races also less prolific, and may permit mankind to solve the population question. This is another reason for desiring a world-government.

Passing from quantity to quality of population, we come to the question of eugenics. We may perhaps assume that, if people grow less superstitious, governments will acquire the right to sterilize those who are not considered desirable as parents. This power will be used, at first, to diminish imbecility, a most desirable object. But probably, in time, opposition to the government will be taken to prove imbecility, so that rebels of all kinds will be sterilized. Epileptics, consumptives, dipsomaniacs and so on will gradually be included; in the end, there will be a tendency to include all who fail to pass the usual school examinations. The result will be to increase the average intelligence; in the long run, it may be greatly increased. But probably the effect upon really exceptional intelligence will be bad. Mr. Micawber, who was Dickens's father, would hardly have been regarded as a desirable parent. How many imbeciles ought to outweigh one Dickens I do not profess to know.

Eugenics has, of course, more ambitious possibilities in a more distant future. It may aim not only at eliminating undesired types, but at increasing desired types. Moral standards may alter so as to make it possible for one man to be the sire of a vast progeny by many different mothers. Prime ministers, bishops, and others whom the state considers desirable may become the fathers of half the next generation. Whether this would be an improvement it is not for me to say, as I have no hope of ever becoming either a bishop or a prime minister.

If we knew enough about heredity to determine, within limits, what sort of population we would have, the matter would of course be in the hands of state officials, presumably elderly medical men. Whether they would really be preferable to Nature I do not feel sure. I suspect that they would breed a subservient population, convenient to rulers but incapable of initiative. However, it may be that I am too sceptical of the wisdom of officials.

The effects of psychology on practical life may in time become very great. Already advertisers in America employ eminent psychologists to instruct them in the technique of producing irrational belief; such men may, when they have grown more proficient, be very useful in persuading the democracy that governments are wise and good. Then, again, there are the psychological tests of intelligence, as applied to recruits for the American army during the war. I am very sceptical of the possibility of testing anything except average intelligence by such methods, and I think that, if they were widely adopted, they would probably lead to many persons of great artistic capacity being classified as morons. The same thing would have happened to some first-rate mathematicians. Specialized ability not infrequently goes with general disability, but this would not be shown by the kind of tests which psychologists recommended to the American government.

More sensational than tests of intelligence is the possibility of controlling the emotional life through the secretions of the ductless glands. It will be possible to make people choleric or timid, strongly or weakly sexed, and so on, as may be desired. Differences of emotional disposition seem to be chiefly due to secretions of the ductless glands, and therefore controllable by injections or by increasing or diminishing the secretions. Assuming an oligarchic organization of society, the state could give to the children of holders of power the disposition required for command, and to the children of the proletariat the disposition required for obedience. Against the injections of the state physicians the most eloquent socialist oratory would be powerless. The only difficulty would be to combine this submissiveness with the necessary ferocity against external enemies; but I do not doubt that official science would be equal to the task.

It may seem as though I had been at once gloomy and frivolous in some of my prognostications. I will end, however, with the serious lesson which seems to me to result. Men sometimes speak as though the progress of science must necessarily be a boon to mankind, but that, I fear, is one of the comfortable nineteenth-century delusions which our more disillusioned age must discard. Science enables the holders of power to realize their purposes more fully than they could otherwise do. If their purposes are good, this is a gain; if they are evil, it is a loss. In the present age, it seems that the purposes of the holders of power are in the main evil, in the sense that they involve a diminution, in the world at large, of the things men are agreed in thinking good. Therefore, at present, science does harm by increasing the power of rulers. Science is no substitute for virtue; the heart is as necessary for a good life as the head.

If men were rational in their conduct, that is to say, if they acted in the way most likely to bring about the ends that they deliberately desire, intelligence would be enough to make the world almost a paradise. In the main, what is in the long run advantageous to one man is also advantageous to another. But men are actuated by passions which distort their view; feeling an impulse to injure others, they persuade themselves that it is to their interest to do so. They will not, therefore, act in the way which is in fact to their own interest unless they are actuated by generous impulses which make them indifferent to their own interest. This is why the heart is as important as the head. By the "heart" I mean, at the moment, the sum-total of kindly impulses. Where they exist, science helps them to be effective; where they are absent, science only makes men more cleverly diabolic. I have no doubt that kindly impulses depend upon the glands. An international secret society of physiologists could bring about the millennium by kidnapping, on a given day, all the rulers of the world, and injecting into their blood some substance which would fill them with benevolence towards their fellow-creatures. Suddenly M. Poincaré would wish well to Ruhr miners, Lord Curzon to Indian nationalists, Mr. Smuts to the natives of what was German South West Africa, the

American government to its political prisoners and its victims in Ellis Island. But, alas, the physiologists would just have to administer the love-philtre to themselves before they would undertake such a task. Otherwise, they would prefer to win titles and fortunes by injecting military ferocity into recruits. And so we come back to the old dilemma: only kindliness can save the world, and even if we know how to produce kindliness we should not do so unless we were already kindly. Failing that, it seems that the solution which the Houynhnms adopted towards theYahoos, namely extermination, is the only one; apparently the Yahoos are bent on applying it to each other.

We may sum up this discussion in a few words. Science has not given men more self-control, more kindliness, or more power of discounting their passions in deciding upon a course of action. It has given communities more power to indulge their collective passions, but, by making society more organic, it has diminished the part played by private passions. Men's collective passions are mainly bad; far the strongest of them are hatred and rivalry directed towards other groups. Therefore at present all that gives men power to indulge their collective passions is bad. That is why science threatens to cause the destruction of our civilization. The only solid hope seems to lie in the possibility of world-wide domination by one group, say the United States, leading to the gradual formation of an orderly economic and political world-government. But perhaps, in view of the sterility of the Roman Empire, the collapse of our civilization would in the end be preferable to this alternative.

INTRODUCTORY NOTE

After passing quickly over ground well known in the nineteenth century (man subduing nature and making a new material environment), Joseph K. Hart in "Power and Culture" presents some fresh thoughts on technology. In making over his material world, man, Hart argued, was making over himself. "History," he wrote, "is the story of the stages through which man and nature have made each other over almost beyond recognition." He also asked a question that haunts Americans a half-century later: "Which is to be master at last, Man or Nature, mind or the machine?"

Hart should attract the interest of a later generation of readers by his awareness of the derivative, or secondary, effects of the machine replacing the skilled worker and artisan. With its cheap, mass-produced goods, the machine of the Industrial Revolution destroyed the old standards of workmanship, taste, and culture. Writing in 1924, Hart also stressed the alienation resulting from man's being "torn loose from former contacts with nature and crowded into the industrial cities, there to live an alien and artificial sort of life." Hart's insistence that the spirit of man needs the nurture of the turbulent sea, the silent, snowy fields, and the wilderness has lost none of its validity.

After a summary discourse on the unfortunate results of man's losing contact with nature, Hart expressed another attitude usually associated with the antitechnology 1960s. The worker had become, Hart believed, the slave of the machine rather than its master: "Nature has had her revenge, for the masses of men, though leading more comfortable lives, have become submerged and dominated by things."

Hart, however, sounded a note of optimism in a conclusion that associates him more with attitudes of the past than those of the 1960s. He saw in the new schemes for electrification a means to disperse industry and to resettle overly dense urban population. He referred to "giant power," a scheme for the economical generation and transmission of electrical power and for rural electrification that was proposed early in the twenties by the governor of Pennsylvania, the conservationist Gifford Pinchot. For Hart, more technology well directed and controlled could solve the problems technology had brought.

Joseph K. Hart, born in Indiana in 1876, was professor of education at Reed College, Portland, Oregon, from 1916 to 1919. From 1920 to 1926, he was an editor for *Survey.* Afterward, he was a professor at the University of Wisconsin, Vanderbilt University, and Teachers College, Columbia University. He wrote a number of interpretive studies on education and society, and among his books are *An Introduction to the Social Studies* (1937) and *Mind in Transition* (1938).

Power and Culture

JOSEPH K. HART

Man is dependent upon nature: "The earth is the mother of all mankind." For food, shelter, clothing, for the materials that help to adorn and beautify, man must go to nature. Though he may not live by bread alone, he cannot live without bread. One need not be an irrational materialist, enamoured of the economic interpretation of history, to admit these facts!

On the other hand, man's most strenuous efforts have been given to the attempt to escape from nature. Nature, primitive and undisciplined, provides precariously for man, keeping him near the margin of subsistence. Below certain levels the earth is a fairly wise mother; above those levels she is stupid and jealous, degrading her children, telling them tales of danger, filling their minds with fear. One need not be an incurable idealist to acknowledge these facts!

An animal, having gathered a few sticks, leaves or straws, for a nest or a lair, is at home with nature, accepting life or death unquestioningly. But man has long been at war with his mother. As an obedient son, using nothing but his bare hands, he was bound to remain a brother of the beasts. Early in childhood, however, he learned a few secrets, such as the use of fire and how to make some simple tools. These gave him the taste of power and a dawning sense of independence. But shortly the jealousy of his mother appeared; she seemed not a true mother, but only step-mother: his hope of happiness lay in outwitting her completely, learning how to control her and make her work for him, thus assuring his escape. Little by little, the meaning of this project has emerged: he must make nature over, on lines nearer to his needs; he must turn nature into a great machine, subject to his control, serving his every desire.

This was an adventure so presumptuous and so perilous

The Survey, Graphic Number 51, (March 1, 1924), 625–628.

that most subsequent ages have called it irreverent, profane. Some have even spoken of it as the "Fall of Man."

His method, slowly evolved, has been "Divide and conquer!" Mastering some bit of nature, he has turned its forces against the rest. The winds have blown his boats to windward. The waters have run up hill. He has poured iron out like water. He has turned night into day. In short, he has decreed his own freedom, and attempted to enter the decree upon the records of time.

Slowly, in the course of his struggles, man has found certain clues to his problem. Otis T. Mason, the American anthropologist, enumerates the five factors with which men must reckon:

> Raw materials, of almost endless variety and usefulness;
>
> Motive powers, from simple muscular energy to the most complex forms of force;
>
> Tools and machinery, in ever-changing types;
>
> Processes of fabrication, simple, complex and compound;
>
> Products, the things sought for, with which man nourishes and supports his life.

At first these products were purely physical: they nourished and supported man's physical life. But in fabrication there was a factor not fully comprehended within the physical; in products there were elements not wholly to be classed as useful; and in man, himself, there was something not dreamed of in the philosophy of nature. Give them time: these are the most important elements in the scene.

Now, in this long struggle, two irritating implications have ever been latent. First, not more has man reshaped nature than nature has reshaped man. Challenging nature, he overcame her in part, and won a momentary control, with freedom. He used that freedom to consolidate his gains and to organize himself into this re-made world. Naively, he assumed that, though he was thus at home in a changed world, he had remained unchanged. He even argued that, however much nature may be made over, "human nature never changes." But the plea is not convincing. The earth has been cultivated in part, and man has become cultured in part. History is the story of the stages through which man and nature have made each other over almost beyond recognition. Man's refusal to admit change is responsible for some great difficulties.

The second of these irritations is a question: Which is to be master, at last, Man or Nature, mind or the machine? History has given various replies to this question. When machines were small and unimposing, men were not greatly cowed by them. But as machines have grown in size and complexity, man has been more and more impressed; until, now, we are not far from worshipping the machine. Our sciences translate us into the likeness of machines. That fact is interesting. If the machine shall reduce us to subjection, Mother Earth will have her revenge: the machine is nature dressed up in modern clothes!

The extent of man's escape from nature may be roughly measured by the surplus he is able to accumulate—the margin of his supplies, actual and potential, above his needs. Surplus is largely a function of the organization of the five factors enumerated above. "Man came to the threshold of civilization," says Morgan, in his Ancient Society, "when he brought about the union of the animal, vegetable and mineral worlds—that is, when he harnessed the ox to an iron plow for the purpose of cultivating the cereals." That organization of power, tools and processes assured a surplus. That surplus enabled man, at least some men, to be free, at least a part of the time. Freedom meant leisure: leisure might mean the discovery of unsuspected values in the world and in human living.

The fundamental factor is power. Materials are everywhere. Tools and fabricating processes and, hence, products will follow upon the coming of power. But power is elusive. Wind and water are fickle, undependable. The strength of animals is dependable; but it is slight, and it requires too much oversight. Power must be plentiful, dependable and requires small oversight. Such power was first found in Slavery.

The slave is the most intelligent application of power to work that has ever been known: he *is* power—intelligent, automotive power. The slave is not efficient; hence, his numbers must be great. But being intelligent, he not only performs work, he can plan and oversee it. Thus, he releases free men completely from nature, providing time for the development of the arts and cultures of life.

These conditions were fulfilled in preeminent measure in Athens, "that point of light in history." A sufficient supply of slave-power freed the citizen from all the stresses of physical existence into a congenial leisure. Freedom revealed the social and moral chaos of the times: barbarian hordes were pressing in to destroy; within himself were areas of impulse, rages and passions that might easily betray him. Having risen above physical nature, he must overcome the barbarian, without and within, and make a world fit for Man.

Led by artists, philosophers and scientists, and by statesmen who, for a time, at least, believed that these have a valid function in society, the Greeks rose above their ancient rages and fears and achieved an ideal world, of serene great beauty, wherein one might meet Fate calmly as became a Man. They became human. They discovered humanity. And they bequeathed it, as a precious treasure, to the ages.

Greek culture was a community enterprise. Slavery made it possible. All the arts and sciences contributed to it. It exalted excellence. Men created beautiful objects, wrote beautiful poetry, trained their bodies to the highest pitch of strength and agility and displayed their prowess in civic contests. They gave to the world a revelation that still haunts the mind. We know, when we are most aware, that any conception of living that neglects beauty is an unworthy, if not a degraded conception.

But there was a defect in Greek life which was inevitably reflected in Greek culture. The community was but partially human: nine-tenths of it was slave. The slave who made Greek culture possible had no share in that culture. Individualists and abstract "humanists" may argue that culture should be above the battle of the classes. The argument gains no support from Athens. Greek culture, compelled to ignore and deny nine-tenths of the people, became remote from life, as it was remote from work; it became intellectual, non-social, fragmentary. A fragment of humanity, however free, can never create more than a fragmentary "humanism." Power, even slave power, laughs at man when he presumes too far upon freedom!

To be sure, the Greeks sought to overcome this defect by a *tour de force:* they ruled the slave outside the limits of the "human." Man in his struggle for control had set himself over against nature. Power is of nature. The slave was a form of power; hence, he belonged to nature, not to humanity. Greek culture was for free men, not for slaves. It was liberal, not servile; for the man of leisure, not for the worker.

If such a distinction had held, Greek culture would have been inclusive of "the humanities." But the Greeks never fully accepted it; and the modern world has denied it. Modern "humanists" have, however, sat at the feet of the Greeks for two thousand years, giving them the homage of sincerest imitation. But the verdict of the modern world seems to be that a "humanity" that could consign a major part of the race outside the bounds of the "human" could, at the best, produce an imperfect, not a true "humanism."

The glory that was Greece failed. Culture is not a veneer upon life: it is of the substance of life. In order to establish an ideal, the Greeks became intellectually dishonest: they degraded some part of their humanity, identified it with nature, and called it "power." Thus they justified their free life. We have the fragments of their culture—nothing else—today.

We pause in the flight of the centuries only long enough to recall that the culture of the Middle Ages was builded upon a quibble. The labor power of the age was the feudal serf. The serf was neither bound not free, neither wholly of nature nor fully human. He was by way of escaping from nature into humanity. He had some share in religion, though not in the civic life. Work was the most effective humanizer. In the later centuries freedom could be achieved by escaping from the land to the cities. The culture of the Middle Ages was largely the work of escaped serfs, who in the cities achieved their moral and artistic enfranchisement as well as their civil liberty. By the thirteenth century, the towns and cities were filled with these freed workers and their guilds, both of the artists and the artisans, were flourishing. Thought of this age centers mostly about the great cathedrals. But beyond those monuments lie centuries of struggle for freedom and control; and all about them are the only slightly less remarkable achievements in every line of artistry. From the thirteenth to the eighteenth century, free workers, whether artist or artisan, were building up standards of integrity in workmanship. Some, still struggling with the past, "wrought in a sad sincerity"; others, feeling the freedom of the future, wrought in a joy hitherto unknown. But each

> . . . "builded better than he knew,—
> The conscious stone to beauty grew!"

And we pay our tribute by prizing their works as "antiques."

Into this world of handcraftsmanship, so human, so artistic, so inefficient, came the Steam Engine, discovering unsuspected reservoirs of power. This power has changed the face of the world—not alone the world of work, but of all the other ranges of human living. These changes constitute the so-called "Industrial Revolution." What have been the characteristics of this age of change?

1. Steam made all other forms of power, wind, water, the slave and the serf, subordinate, and promised quickly to make them obsolete. That is to say, steam organized about itself the industries of the world.
2. Steam tore people loose from their local communities and began to crowd them into narrow areas around new centers of industry. It has successively torn people of all stages of culture loose from their old rootages in local groups and gathered them into polygot centers of industry and commerce. Steam has been, and is, the great centralizer.
3. Steam has supplied the world with unlimited products for nourishing and adorning life: an incredible range of useful and useless implements, weapons, tools, gewgaws, impedimenta; machines of transportation; and, by means of a variation, the gas engine, with the automobile and the flying machine. These, by their very nature the tools and the *means* of life, have charmed and bewildered us until we have transformed them into the *meanings* and the ends of life.
4. Hence, steam has torn us free from old standards of workmanship, taste and culture. By giving us cheap, machine-made articles in unlimited quantities in place of the older hand-made objects, steam has made the world more comfortable; but at the price of substituting display and exaggeration for use and beauty. Inevitably, the nineteenth century was an epoch of vulgar comfort.

Within a century, at least half the people of the world felt these effects in some degree. In the western world, at least one-third of the population has been torn loose from former contacts with nature and crowded into the industrial cities, there to live an alien and artificial sort of life. The industrial city rests upon the steam engine. "Large-scale organization," of industry and of living has been the keynote of the industrial revolution.

But man's inner life is responsive to his environment. This is "adaptation." Now, since environment in the modern city is almost completely artificial, man has necessarily become artificial, also. His culture has become artificial. He has made himself too free from nature. He has shut out the stars with his roofs and his smoke. He no longer hears the breaking of waves over deep seas; no longer fights with the wilderness far on the frontiers; no longer follows the aurora over the silent snow-fields. He has ceased to renew his spiritual life at the ancient springs. Books tell of men and women who once did those things; but the books are scarcely credible. Men go, it is true, at times, to the mountains or the sea, for the purpose of escaping from spiritual dyspepsia—in order that they may, once more, enjoy the feast the city offers. And what is that feast?

Bacon has told us how the scholastics, turning ever inward for the materials of their dialectic, were like spiders that spin endless threads of disputation out of their own bodies. We have been moving in similar directions. When the mind of man loses contact with nature, it turns in upon itself, and spins out of its own memories, endless repetitions, endless monotonies. Its art becomes superficial and clever: endless variations upon the same literary theme; machine-made music; pictures by wholesale. Lacking ideas, it writes poems with the supreme distinction that each line begins with a small letter.

Even so, men cannot live in this way forever. Bored by it, but having no other escape, they revert to primitive nature, to as much of reality as remains to them: to undisciplined instinct; to movies that show "he-men"; to jazz and dances that exaggerate sex-motives; to novels that reek with decadent sex-recitals; to pugilistic encounters that smell of blood; to court trials that display sadistic experiences; to theatricals that "exalt the human form."

Meanwhile governmental and industrial leaders, protecting their own freedom, after a fashion, by trips to Palm Beach or to Europe, talk glibly of the "advantages of the machine era." It has given us, they say, comforts beyond the dreams of avarice in other eras: health, prosperity, long life; a standard elementary education for all; unimpeachable patriotism; seats in the park; and athletic contests more thrilling than any since the days of the circus in Rome. And our culture leaders, admitting as glibly that there can be no hope of an indigenous culture in our machine civilization, happily tell us that what we need, and "all that we need" is an importation of the "humanities" of the Greeks. "We have utility; the Greeks had 'humanity': add them together and secure the finest civilization possible to men!"

But some are not wholly convinced by these reputed virtues of the machine age: freedom by proxy and culture by addition. The sense of loss is too great:

We have lost contact with nature—the contact that gave to man his first challenges, his first joy of battle, his first sense of victory.

We have lost that neighborliness which was characteristic of the older community, when men lived in homes and worked with their hands. The steam engine first undermined that community, and the automobile has completed its destruction.

We have lost practically all of the integrity of our old craftsmanship. The machine is not interested in integrity: only in form. Both the artist and the artisan have suffered spiritual dislocation. The artisan now works, dispiritedly, for the machine; and the artist, competing with the machine, too often sells his soul to feed his body.

We have lost practically all control of our destinies. We work when the machine works; we do what the machine commands; we use the products the machine turns out. We are educated to work with the machine and to use machine-made products.

In the centuries of free workmanship, especially in pioneer America, men were moving slowly toward a finer humanity, a real community, in which every individual might find himself at home. They dreamed of liberty and fraternity. Perhaps it was a fool's dream! At any rate, the steam engine, in building our industrial cities, has cut us, more or less sharply, into two groups once more: the owners and the workers. The owners, as free and independent centers of control, make up humanity; the workers, as attachments to the machine, are not sure where they belong!

They are not slaves, bound to the wheel of labor: they can always give up their jobs—at least one at a time. In religion and politics they boast equality with the owners. They are schooled at the expense of the community. But in economics they are still classed as "labor power," and they are dealt with as if they were something that humanity must control in order to maintain its own precariously achieved freedom.

Since about 1890, electricity has been more and more applied, subordinately, to the steam-driven machine, making that machine more completely automatic. (Electricity has not been, at least until quite recently, a power in its own right; it has been a helper of other power.) Working with automatic machines, the worker has grown more automatic: an "iron man," a "robot." His task can be learned in a few hours, or days, at most; and once learned, it can be changed only with the greatest difficulty. "It never was so easy before for a simpleton to live!" Nature has had her revenge: man's long struggle with her has come to this, that for the masses of men, while their physical lives are far more comfortable, less precarious, than was the case with primitive men, their mental and cultural lives are more completely submerged in "things" than has been the case since the first few awkward upward steps were taken in the primitive wilderness.

The culture of the Greeks grew, as we have seen, out of a great stress: the struggle of noble spirits against the overpowering Fates; the struggle for Order in the midst of an all-devouring chaos, for a Mind that should rise above non-mind and find or give a meaning to existence. They failed; but that was what they sought!

Today we have reached the sublime belief that the Machine is the nearest approach to Reason and Mind that we shall ever reach: the Machine stands between us and every sort of chaos: it feeds us, clothes us and educates us; it fights our battles for us and organizes our peace. The Machine is the last word in cosmic progress. We have substituted it for the spirit that once was in us; we have made ourselves over, in our psychologies, on the model of the Machine: we have lost our souls for it; some say we have even lost our minds! Is there no escape for the race from these untoward, tragic happenings?

The industrial city is too unhuman to be the home of the human spirit. If, for Plato, Athens was too large, what shall be said of our modern aggregates of shifting, drifting men? It is true that the great city has become the center of the greatest stimulations the world has ever known. But these stimulations are practically all upon the periphery of life: they do not reach the center. Such peripheral stimulations make for cleverness and smartness: for the literature of Gertrude Stein, which has no "message," only "suggestions"; for arts whose boast is that they have no meaning.

But what shall the soul do that cannot nourish itself on words; that longs to find the ancient springs where it may drink long at the fountain of living waters? Where shall it find that "silence and slow time," of which, according to Keats, culture is the foster-child? Are all such questions infantile survivals which the "manhood of humanity" should have long escaped? Should things have no meaning? If things should come to have meaning would that fact challenge the dominance of the Machine in human living?

The question has been raised. Men are in revolt against an industry in subordination to which, they "dig the ditch, in order to get money, with which to buy food, so that they can have enough strength to dig the ditch!" The Machine has taken on the form of Fate, remorseless Fate. Two things men want today—to wit: contact with the earth once more, and more neighborly contacts with their human kind! And these two desires seem not so unattainable today as they seemed five years ago!

Centralization has claimed everything for a century: the results are apparent on every hand. But the reign of steam approaches its end: a new stage in the industrial revolution comes on. Electric power, breaking away from its servitude to steam, is becoming independent. Electricity is a decentralizing form of power: it runs out over distributing lines and subdivides to all the minutiae of life and need. Working with it, men may feel the thrill of control and freedom once again.

Life need no longer be subordinated to steam: Industry can

be decentralized—the smaller community can be regained, with its old humanities. The mechanisms of such decentralization now wait man's use: has he the courage to make the world he needs? He could not control the past, for he could not foresee its direction. But now the future lies open before man, as it did in the day when Joshua said to Israel: "Behold, I have set before you life and good, death and evil: choose ye this day which ye will serve!"

Humanity has no spiritual future save in the fight for that economic and social freedom within which the mind can be free. Giant power, under public control, with power distributed to all on equal terms, offers economic freedom to humanity, the hope of communities within which intellectual freedom can be realized and the culture of the spirit will seem possible.

Such decentralization of living will tend to regenerate our culture by releasing it from the city's hot-houses, where it attains a superficial brilliance, and restoring it to its native rootage in reality. In the reinvigorated small community, the free mind will become creative; and schools, within which freer minds may develop, will appear once more.

Holders of vested interests in our present economic order will oppose these developments; and properly so, for salvation must not be too easy. Humanists whose culture is an imitation of the Greeks will also oppose them: a humanism indigenous to our soil would not be to their liking.

For mankind, these are crucial times. Wishing can do little. But thinking can lay hold upon the materials of the future and make a world in which humanity will be freed and enfranchised; or a world in which humanity will lose itself under the mazes of economic mechanisms.

This is the Day of Choosing: We stand, today, where the Greeks once stood: face to face with Fate. We have Power beyond their dreams of power: power that indisputably belongs in the realm of nature, the proper use of which need not degrade a single human being. We can see the Fates at work. We can build communities upon the foundations of great but decentralized power, we can build small communities where life and culture can be rooted in normal relationships. We can provide the materials out of which men can make for themselves the manner of life they prefer. Or, we can surrender to the control of the greater machine, permit electricity to make permanent what the steam-engine began, be happy in the roar of industry and lose all our sense of freedom, justice and beauty. "The history of the world is the world's judgment day!"

INTRODUCTORY NOTE

The ambivalence of the postwar era resulted from the horrors of war fought with the power of modern technology and science, the deterioration of the industrial environment, the thoughtless exploitation of natural resources for technological processes, and the awareness of other contrasts between what nineteenth-century Americans had envisaged as the fruits of applied science and technology and what the twentieth century was experiencing. There was affluence and power, but—as Arthur D. Little observed in his 1924 address, "The Fifth Estate"—there was also upheaval and unrest, "unnecessary illness, crushing taxation," inefficient government, and "the appalling waste of effort, material and resources." Like the serfs of Russia, who would not blame the czar but his ministers for the woes of society, Little and many other Americans blamed the politicians who, they believed, misused technology in governing society. He also spoke harshly of the uneducated and insensitive masses of people who supinely accepted ignorant and corrupt leaders.

To rectify the situation, to place modern science-based society on the rails again, Little called for the leadership of the "fifth estate." It would cultivate science and technology so that society could reap the harvest foreseen in the nineteenth century. The fifth estate, a small group in the 1920s, was the community of science and scholarship. In arguing his case, Little asserted that the fifth estate's competence extended beyond thought to action. Frankly elitist, he called for the ignorant to accept the authority of the learned and of those who knew the "divinity" of science. He advocated planning social and economic change by having the fifth estate formulate national policies and solve government problems. (He also advocated selective breeding of animals and men.) His call for planning would often be echoed in the twenties and thirties.

Arthur D. Little (1863–1935) was a distinguished industrial chemist known especially for the excellence of the independent industrial research laboratory, Arthur D. Little, Inc., that he founded, and for the analytical clarity of "unit operations," a concept of chemical processes that he formulated. But he also enjoyed a national reputation for his literary interests and his writing ability and became a leading spokesman for scientists on professional and social questions. A collection of his essays was published under the title *The Handwriting on the Wall* (1928).

The Fifth Estate

ARTHUR D. LITTLE

I

Benjamin Franklin was not perhaps in all respects a paragon, but he was unquestionably a polygon—a plain figure with many sides and angles. There were not enough buttons on his black coat to tell off the multifarious aspects in which his complex personality was presented to the world. He was craftsman and tradesman; philosopher and publicist; diplomat, statesman, and patriot. And he was, withal, a very human being. What concerns us particularly on this occasion is the fact that he was at once philosopher and man of affairs. His remarkable career should refute forever the fallacy which, unfortunately, still is current, that the man of science is temperamentally unfitted for the practical business of life.

At the time when Franklin was in England the British Parliament was assumed to be composed of representatives of three estates: the lords spiritual, the lords temporal, and the commons; but Edmund Burke, pointing to the Reporters' Gallery, said, "There sits a *Fourth Estate*, more important far than they all." No one at all familiar with the ubiquitous influence and all-pervading power of the press would to-day question the validity of Burke's appraisal. Even then, however, there was present in England, in the person of Benjamin Franklin, a prototype and exemplar of the membership of a *Fifth Estate*, an estate destined to play an even greater part than its predecessors in the remaking of the world.

This Fifth Estate is composed of those having the simplicity to wonder, the ability to question, the power to generalize, the capacity to apply. It is, in short, the company of thinkers, workers, expounders, and practitioners upon which

---

The Atlantic Monthly, 134 (December 1924), 771–781. Originally an address delivered at the Franklin Institute Centenary, 1924. Reprinted by permission.

the world is absolutely dependent for the preservation and advancement of that organized knowledge which we call Science. It is their seeing eye that discloses, as Carlyle said, "the inner harmony of things; what Nature meant." It is they who bring the power and the fruits of knowledge to the multitude who are content to go through life without thinking and without questioning, who accept fire and the hatching of an egg, the attraction of a feather by a bit of amber, and the stars in their courses, as a fish accepts the ocean.

The curious deterioration to which words are subject has left us with no term in good repute and common usage by which the members of the Fifth Estate may properly be characterized. Sophists are no longer distinguished for wisdom, they are now fallacious reasoners. Philosophers, who once claimed all knowledge for their province, are now content with speculative metaphysics. Scholars have become pupils. The absent-minded and myopic professor is a standardized property of the stage and screen. The expert, if not under a cloud, is at least standing in the shade. In Boston one hesitates to call a professional man a scientist—he may be a Presbyterian; and a "sage," as an anonymous writer has pointed out, "calls up in the average mind the picture of something gray and pedantic, if not green and aromatic." Let us, therefore, for a time at least, escape these derogations and identify ourselves as members of the Fifth Estate. Although the brotherhood of the Estate is open to all the world, its effective membership nowhere comprises more than an insignificant proportion of the population. Two hundred and fifty constitute the membership of the National Academy of Sciences. The latest edition of *American Men of Science* includes only about 9500 names. The number is expanded to 12,000 on the roll of the American Association for the Advancement of Science. Although gathered from all countries,—and though chemistry is one of the most active and inclusive sciences,—the chemical papers, books, and patents reviewed in *Chemical Abstracts* in 1923 were the product of about 22,000 workers. One may hazard the estimate that there are not in all the world 100,000 persons whose creative effort is responsible for the advancement of science.

The studies of Cattell indicate that in America, at least,

the great majority of men of science come from the so-called middle and upper classes, or precisely those sections of society which, in Russia, have been practically exterminated in the name of the new Social Justice. In about two thirds of Cattell's reported cases both parents were American-born, while the fathers of nearly one half were professional men. Seventy-five per cent depend upon the universities for support; from which we may assume that the burden of the higher surtaxes does not bear heavily upon the Fifth Estate.

In proportion to population the cities have produced twice as many scientific men as the country, but how many hearts "once pregnant with celestial fire" repose in country churchyards because of lack of opportunity and absence of the stimulus of contact cannot, of course, be known, nor can we tell how many brains, competent and well equipped to penetrate the mysteries of nature, the war has cost the world.

Initiative is one of the rarest mental qualities, yet without it progress is impossible. Its combination with the scientific imagination and command of fact is still rarer and more precious. Since comparatively few of those who study science develop the capacity to extend its borders, the cost of a man competent to advance science has been estimated at $500,000 and his value to the community set at a far greater figure. Full membership in the Fifth Estate thus seems to involve the highest initiation-fee on record. It is a figure disconcerting to the candidate, but as Wiggam has finely said: "Only genius can create science, but the humblest man can be taught its spirit. He can learn to face truth."

That the Fifth Estate is not better appreciated or always understood by the world at large is not surprising. In their endeavors to secure accuracy of definition and expression its members have evolved a preposterous and terrifying language of their own. It is not ideally adapted to the interchange of confidences in ordinary human intercourse. It does not lend itself to poetry. "Ladybird, ladybird, fly away home" becomes impossible when one is forced to address the prettily spotted beetle as *Coccinella dipunctata*. A primrose by the river's brim is much more than a yellow primrose to the botanist: it is a specimen of *Primula vulgaris*. The organic chemist produces a new synthetic product in a mass of pilular dimensions and bestows upon it a name that would slow up Arcturus. Nothing but static interference can account for the terms of radiotelephony.

If knowledge is to be humanized it must first be translated.

Dewar has said that the chief object of the training of a chemist is to produce an attitude of mind. It should be the object of all education to produce the scientific attitude toward truth. We may even agree with Robinson that "of all human ambitions an open mind, eagerly expectant of new discoveries and ready to remould conviction in the light of added knowledge and dispelled ignorances and misapprehensions, is the noblest, the rarest, and the most difficult to achieve."

Carlyle says, "The degree of vision that dwells in a man is a correct measure of the man." And President Coolidge has been quoted as saying in a recent interview:—

> Everything flows from the application of trained intelligence, and invested capital is the result of brains. . . . The man of trained intelligence is a public asset. . . . We go forward only through the trained intelligence of individuals, but we, not the individuals, are the beneficiaries of that trained intelligence. In the very nature of things we cannot all have the training, but we can all have the benefits.

Now vision, a trained intelligence, and an open mind are the qualities which characterize all those who are worthy of membership in the Fifth Estate. They are qualities which the many-sided Franklin possessed in exceptionally high degree.

II

Among all the activities with which his busy life was crowded Franklin undoubtedly found his greatest pleasure in the pursuit of science, and in that pursuit he followed the eclectic method. At a time when nearly everything awaited explanation his focused attention ranged like a searchlight over many fields. He observed the movement of winds and developed a theory of storms. He considered ventilation and the causes of smoky chimneys and proceeded to invent new stoves. He introduced the Gulf Stream to Falmouth skippers and demonstrated the calming effect of oil on turbulent seas to officers of the British Navy at Portsmouth. From earthquakes he turned to the heat-absorption of colored cloths and the fertilizing properties of gypsum. He wrote on sun spots and meteors; waterspouts, tides, and sound. The kite, which for centuries had been the toy of boys, became in Franklin's hands a scientific instrument, the means to a great discovery. That its significance is, even now, not universally appreciated is shown by the recent answer of a schoolboy, "Lightning differs from electricity because you don't have to pay for lightning." To Franklin, as the child of every man knows, we owe our initial conceptions of positive and negative electricity, and he was the first to suggest that the aurora is an electrical phenomenon.

The gregariousness, which is a prominent characteristic of the Fifth Estate, found early expression in Franklin. He formed the Junta, a club for the discussion of morals, politics, and natural philosophy, and in 1743 drew up a proposal for the organization of the American Philosophical Society, of which later he became president. He established a wide acquaintance and cemented many firm friendships among the foremost scientific men of France and England, by whom he was received on equal terms. In 1753 he was awarded the Copley medal of the Royal Society for his discoveries in electricity and, on his leaving England, David Hume wrote:

> I am sorry that you intend soon to leave our hemisphere. America has sent us many good things,—gold, silver, sugar, tobacco, indigo,—but you are the first philosopher and, indeed, the first great man of letters for whom we are beholden to her.

The professional spirit which animates the Fifth Estate is essentially one of service. Its compelling urge in the search for truth springs from the conviction that the truth shall make men free. That spirit finds complete expression in Franklin's statement, "I have no private interest in the reception of my inventions by the world, having never made, nor proposed to make, the least profit by any of them." This impersonal relation to the children of his brain was indeed carried by him to an extent which ordinary human nature would find hard to emulate. "I have," he writes, "never entered into any controversy in support of my philosophical opinions; I leave them to take their chance in the world. If they are right, truth and experience will support them; if wrong, they ought to be refuted and rejected."

There is, nevertheless, a place for militancy in science. The world needs a Huxley for every Bryan.

Franklin was a man of science, but his career proclaims that it is possible to be a man of science and much more besides. Science was made for life, and life is more than science. Art in its fullest expression may touch deeper springs, human relations and affections may bring richer rewards, and public affairs may make a more imperious claim. With Franklin as their prototype the members of the Fifth Estate may well strive to emulate his devotion to the public service and his constructive interest in human affairs.

Error and misconception have a feline tenacity of hold upon life, and the Fifth Estate, though richly endowed with latent executive capacity, is still, in popular opinion, regarded as equipped for thought rather than for action. The practical man, busily engaged in repeating the errors of his forefathers, has little time and less consideration for the distracting theories and disconcerting facts of the man of science. Yet who, among the men of action, is more intensely and truly practical than Carty, Baekeland, Reese, or Whitaker? Where shall one find a firmer grasp on the details of business than that possessed by E. W. Rice, Jr., Gerard Swope, or Dr. Nichols? What quality caused the young director of a research laboratory to find himself responsible for the production of gas masks to protect four million fighting men? In a time of dire emergency it was a professor of chemistry who organized the great Edgewood Arsenal and developed the means and methods and the trained personnel required to supply munitions for a new type of warfare. It was not to a statesman or a business man or a great manufacturer that the Allies entrusted the supreme command. It was to a teacher in a French military school. The range and value of their public service obscures the fact that Charles W. Eliot was a professor of chemistry and that Hoover is an engineer. The League of Nations is the child of a schoolmaster.

Numerically the Fifth Estate has always been feeble and insignificant. Its total membership at any time could be housed comfortably in a third-rate city. No politician makes a promise or invents a phrase to attract its scattered and ineffective vote. Rarely do its members sit in Congress; when they do they sit in the gallery.

With less political influence than the sparse population of Nevada, the Fifth Estate has recast civilization through its study and application of "the great and fundamental facts of Nature and the laws of her operation." It has opened out the heavens to depths beyond imagination, weighed remote suns, and analyzed them by light which left them before the dawn of history. It has moved the earth from the centre of the universe to its proper place within the cosmos. It has extended the horizon of the mind until its sweep includes the 30,000 suns within the wisp of smoke in the constellation Hercules and the electrons in their orbits within the atom. It has read the sermons in the rocks, revealed man's place in nature, disclosed the stupendous complexity of simple things, and hinted at the underlying unity of all.

Because of this new breadth of vision, this lifting of the corner of the veil, this new insight into the hidden meaning of the things about him, the mind of man, cramped for ages by taboos and bound by superstition, is emerging into freedom: into a new world, rich in promise, and of surpassing interest and wonder.

Man brought nothing into the world and through long and painful ages he added little to that nothing: a club, an axe of stone, a pebble in a sling, some skins of beasts, a rubbing of sticks for a fire. He might labor, but to what avail? Even to-day the South American Indian works incessantly, yet his labor produces little more than heaps of stones. To those who would have us believe that all wealth is produced by labor the Fifth Estate replies, "Wealth is the product of brains, and labor is productive only as it is guided by intelligence."

Science is the great emancipator of Labor. Bagehot has somewhere said, perhaps in *Physics and Politics*, that during the early stages of civilization slavery was essential to progress because only through the enforced labor of the many could the few have leisure to think. To-day, in the United States, the supply of available energy is equivalent to sixty manpower for every man, woman, and child. There is now leisure for all to think, but the millions prefer the movies.

It is not Labor, but the trained intelligence of the Fifth Estate which has endowed man with his present control of stupendous forces. It has solved problems that for ages have hindered and beset mankind. It has revealed great stores of raw materials, synthesized scores of thousands of new compounds, furnished the fundamental data which find embodiment in machines and processes and in those agencies of transportation and communication that have made of the world a neighborhood. It has enabled man effectively to combat disease, added years to the average life, and made it better worth the living.

III

Benjamin Franklin died in 1790—one hundred and thirty-four years ago. Could he return to make appraisal, what wonders would confront his astonished vision, what triumphs of the Fifth Estate compel his admiration!

Electricity, which to his contemporaries was little more than an obscure force, the curious manifestations of which might supply an evening's entertainment, has become the structural basis of the universe. The atom of Democritus is now a microcosm, vibrant with energy that glows in the white light of the electric lamps, which have replaced the tallow dip. In place of the electrophorus and the charges of the Leyden jar he would find in our own country alone twenty-seven million horsepower driving generators in thousands of stations from which electric energy is distributed to our homes and factories and transportation lines to perform innumerable services. Imagine, if you can, the stunning impact of the impressions that would crowd the day of his return. With what amazement would he converse over a wire from Philadelphia to San Francisco or hear a voice transmitted through the ether from a point halfway around the world. So commonplace a thing as a street car would leave him open-mouthed with wonder, which might well increase at sight of an electric locomotive, hauling its hundreds of tons of freight.

In great industrial plants he would find electricity driving machines of an intricacy, precision, and productive power beyond the imagination of his generation, or at work in decomposing cells, and in the heart of glowing furnaces fashioning new products. In university and corporation laboratories would be revealed to him the marvel of the X-rays, photography, the fascinating world of the microscope, balances weighing 1/100,000th of a milligramme, the spectroscope, and all those instruments of precision and research which are the tools of the Fifth Estate. Elements unknown to him would be placed in his hand; fascinating experiments performed to demonstrate properties and relationships beyond his dream. The air, which he studied with reference to winds, combustion, and ventilation, would be reduced before him to a liquid as obvious as water, though boiling on a cake of ice.

Where once the postboy and the post chaise were familiar he would find our roads crowded with automotive vehicles and the country gridironed by the railways. Did he wish to send a letter across the continent, he would have only to commit it to the air mail to ensure its arrival in thirty-six hours. Were he called upon to revisit England, there would be no ten weeks' voyage in a sailing packet, but the speed and luxury of a 50,000-ton liner, oil-fired and turbine-driven. At Portsmouth, where he calmed the waves with oil, he would find, instead of wooden frigates and smooth-bore cannon, submarines and armored superdreadnoughts, a single gun of which could sink the entire British Navy as he knew it. Did he wish to proceed to Paris? He would have only to take passage in an airplane.

The gardeners Franklin knew grew peas for pleasure or profit. Mendel grew them and established the laws of heredity. Farming, which was a wholly empirical occupation, is now the special concern of a great governmental department devoted to the development of scientific agriculture. Here Franklin would learn of soil analysis and seed selection, of hardier and more prolific varieties of plants, of better breeds of animals, of methods of control of such virulent diseases as splenic fever, anthrax, hog cholera, and bovine tuberculosis. He would find his own experiments with gypsum extended to cover the whole field of chemical fertilizers, the air itself converted into an inexhaustible reservoir of plant food, and the efficiency of farm labor multiplied many times by ingenious agricultural machines.

He would find household economics revolutionized: the town pump replaced by running water; electricity a servant in the house; the food supply broadened and stabilized; domestic drudgery assumed by laundry, bakery, and factory; tasteful clothing within the reach of all; transportation and amusement for the multitude, and the history of yesterday sold for a penny; innumerable new industries, based on the findings of the laboratory, now offering means of decent livelihood to millions, opening careers to thousands.

In great hospitals, permeated with the scientific spirit and equipped with many new and strange devices for the alleviation of human suffering, he would hear of the incalculable benefits which medical and surgical science have conferred upon mankind. He would see the portraits and listen to the story of Pasteur and Lister and Loeb and Ehrlich. We know to-day with what joy and relief the world would welcome a veritable cure for cancer, but we can little realize the emotion with which one like Franklin would learn in a single afternoon of the germ theory of disease, of preventive serums, of antisepsis, of chemotherapy, of the marvelous complexity of the blood stream and the extraordinary influence and potency of the secretions of the ductless glands. What appraisal would he make of the service to humanity which, in little more than a generation, has mitigated the horrors of surgery by the blessings of anaesthesia and antisepsis, which has controlled rabies, yellow fever, typhoid fever, tetanus, which is stamping out tuberculosis, curing leprosy, and providing specifics for other scourges of the race? What values would he put on insulin, thyroxin, adrenalin? The physician is no longer compelled to rely on herbs and simples and drastic mineral compounds of doubtful value and uncertain action. Compounds of extraordinary potency, isolated or synthesized by the chemist, are now available to allay pain, correct disorders, prolong life, and even to restore mentality and character.

With contributions to their credit which have so enriched and stimulated the intellectual life; which have brought the peoples of the earth together into closer touch than English shires once were; which have revolutionized industry, enlarged the opportunity of the average man, and added so greatly to his comfort and well-being, we may reasonably inquire, "What are the recompenses of the Fifth Estate?"

On the material side they have almost invariably been curiously inadequate and meagre. It is incomparably more profitable to draw the Gumps for a comic supplement than to write the *Origin of Species*. There is more money in chewing-gum than in relativity. Lobsters and limousines are acquired far more rapidly by the skillful thrower of custard pies in a moving-picture studio than by the no less skillful demonstrator of the projection of electrons. The gate receipts of an international prize-fight would support a university faculty for a year.

One may recall that Lavoisier was guillotined by a republic that "had no need of chemists"; that Priestley was driven from his sacked and devastated home; that Leblanc, after giving the world cheap alkali, died in a French poorhouse; that Langley was crushed by ridicule and chagrin in his last days. A month before the war who could have believed that within a few years the Fifth Estate in Russia would be utterly destroyed and in Germany and Austria existing at the very edge of starvation? What has happened there may happen again elsewhere if the intelligence of the world does not assume and hold its proper place in the direction of national and world affairs.

In the preface to his recent *Lehrbuch der Photochemie* Professor Plotnikow has written:

> Home and property were pillaged by bands of idle Russians who used my library for cigarette papers. Hunger, misery, want, and personal insecurity, often approaching fear for my life, were the constant accompaniment of my labors.

One is reminded that Carlyle, on the authority of Richter, says:

> In the island of Sumatra there is a kind of "Light-chafers," large fireflies, which people stick upon spits, and illuminate the ways with at night. Persons of condition can thus travel with a pleasant radiance, which they much admire. Great honor to the fireflies, But—!

It is not becoming that the world expect the light to shine indefinitely when carrying a lantern is often less remunerative than carrying a hod. The money and the years of study required for special training are not recognized as invested capital, and the return from a decade of research is often taxed as the income of a year. Professorial salaries move forward as slowly as a glacier, but they seldom leave a terminal moraine. Yet teaching is our most important business; for a failure to pass on for a single generation the painfully accumulated knowledge of the race would return the world to barbarism.

Though material wealth is rarely acquired by the Fifth Estate, they have the riches of the royal man, defined by Emerson as "he who knows what sweets and virtues are in the ground, the waters, the plants, the heavens, and how to come at these enchantments." Their wealth is in the Kingdom of the Mind. It is inalienable and tax-exempt. It may be shared and yet retained.

A recent survey by a national magazine would seem to indicate that the majority of men have drifted into their vocations with little effort of selection and that a very large proportion ultimately regret their choice. This is seldom true of members of the Fifth Estate. Theirs is a true vocation, a calling and election. It brings intellectual satisfactions more precious than fine gold. They live in a world where common things assume a beauty and a meaning veiled from other eyes; a world where revelation follows skillful questioning and where wonder grows with knowledge. Together they share the interests, the communion of spirit, the labors and the triumphs of the fraternity of Science. The Law of Diminishing Returns exerts a control from which there is no escape in agriculture, industry, and business. Research alone is beyond the twelve-mile limit of its inhibitions.

If the heavens declare the glory of God that glory is surely made more manifest by telescope and spectroscope. If the whirling nebulæ and the stars in their courses reveal Omnipotence, so do the electrons in their orbits reveal His presence in universes brought into being by the striking of a match. The laboratory may be a temple as truly as the church. The laws of Nature are the Will of God, their discovery is a revelation as valid as that of Sinai, and by their observance only can man hope to come into harmony with the universe and with himself.

There has been a general and ready acceptance by the world of the material benefits of science, while its contributions to sociology and ethics are as generally ignored as guides to human conduct. Yet science proclaims new commandments as inflexible as those engraved on stone, and furnishes what Wiggam has reverently termed "the true technology of the Will of God."

Science has so drawn the world together and so rapidly remoulded civilization that the social structure is now strained at many points. Statecraft and politics, law and custom, lack the plasticity of science and are now in imperfect contact with the contours of their new environment. The result, as events have shown, is friction and confusion. Though our civilization is based on science, the scientific method has little place in the making of our laws. Office does not seek the man in the laboratory, and candidates are not pictured as engaged in any activity that might suggest a superior intelligence. They are shown milking cows, pitching hay in new blue overalls, or helping with the family washing. Recently, in the senate of a New England state, there was presented the edifying spectacle of the presiding officer, being shaved by a barber, called to the rostrum, while senators were reading the encyclopædia into the record. To expedite further the public business sundry members of the chamber were presently gassed with bromine. Does not this suggest that a few chemists might with advantage be distributed among our legislative bodies?

It is claimed that fifty per cent of the members of state legislatures in America have never been through high school and that only one in seven has been through college. We see in the ranks of science knowledge without power and in politics power without knowledge. An electorate, which regards itself as free, listens to the broadest noise of manufactured demonstrations and is blind to the obvious mechanics of synthetic bedlam. The result is too often government by gullibility, propaganda, catchwords, and slogans, instead of government by law based on facts, principles, intelligence, and good will.

As President Stanley Hall once said, "Man has not yet demonstrated that he can remain permanently civilized." Many thoughtful people have been led to question the ultimate effect of science upon civilization. We all recognize the utility of matches, but we keep them away from children. Meanwhile, science puts dynamite and TNT, poison gas, airplanes, and motor cars at the disposal of criminals and leaders of the mob. Bertrand Russell, in *Icarus,* sees in science the ultimate destroyer. Haldane, in *Dædalus,* visualized it as the stern and vigorous chastener and corrector which will ultimately save the race and usher in the new day of light and reason.

"Knowledge comes, but wisdom lingers," and democracy levels down as well as up. Even in Boston cigars have replaced books on a corner famous for a century of literary associations. The world is wrong because few men can think. It will not be made right until those who cannot think trust those who can. When its foundations are so obviously out of joint humanity still clings tenaciously to fossilized precepts and opinions and is as resentful of suggested change as in the days of Galileo. Despite the pressure of new ideas, education must still, to be acceptable, follow old conventional lines.

IV

Let us not deceive ourselves. Human life is still a hard and fearsome thing. Mankind is required to maintain existence in a world in which, as Kipling has said, "any horror is credible." More than a hundred years ago De Quincey wrote, "We can die, but which of us, knowing, as some of us do, what is human life, could, were he consciously called upon to do it, face, without shuddering, the hour of birth?" But little more than yesterday Henry Adams closed his *Education* with the expression of the hope that perhaps some day, for the first time since man began his education among the carnivores, he would find a world that sensitive and timid natures could regard without a shudder.

Everywhere there is upheaval and unrest. "The machine," to quote Dr. Elton Mayo, "runs to an accompaniment of human reverie, human pessimism, and sense of defeat."

We are everywhere overburdened by unnecessary illness, crushing taxation, extravagant and inefficient governments, huge expenditures for trivialities, and the appalling waste of effort, material, and resources. We are hampered by class suspicion and misunderstanding, racial antagonisms, the inhibitions of organized labor, and the lack of imagination in high places. Life in general is on a low cultural plane and bound by custom and tradition.

One hundred years of science have failed to satisfy the cravings of humanity. Chesterton finds science "a thing on the outskirts of human life—it has nothing to do with the centre of human life at all." We do not, of course, agree with him, but we must still meet the challenge of John Jay Chapman, who declares: "Science, which filled the air with so large a bray, is really a branch of domestic convenience, a department for the study of traction, cookery, and wiring. The prophet-scientists have lived up to none of their prospectuses." The fault, however, as Wiggam points out is not with science, nor with the scientists. It is with those who "have mainly used the immense spiritual enterprise of science to secure five-cent fares, high wages, and low freight rates," when it should have "ushered in a new humanism."

Thus we still encourage race deterioration, still carry the burden of the unfit, still cultivate national antipathies, still are breeding from poor stock, and witnessing with equanimity the suppression of the best.

The history of aristocracies, feudalism, the Church, the guilds, and the soviets has amply demonstrated that no one class possesses the qualities required for the government of all classes, and we cannot claim them for the Fifth Estate. We can, however, claim with full assurance that the Fifth Estate possesses many qualities, now practically ignored, which could be utilized in government to the incalculable advantage of us all. Its knowledge of material facts, of natural and economic laws, of the factors governing race development and human relations; its imagination, vision, and its open mind, should be brought to bear effectively in the formulation of national policies and the solution of governmental problems. There is an alternative before us, which has recently been defined with somewhat surprising frankness by Warren S. Stone, President of the Brotherhood of Locomotive Engineers, perhaps the most conservative of the labor unions, Mr. Stone says:—

> But until labor, in the inclusive sense in which I am using it, secures control of legislative and executive branches of the national and state governments, and through control of the executive branch secures control of the judiciary, labor is in continuous peril of seeing its gains wiped out and its progress retarded by hostile legislation or unfriendly court decisions.

Our countrymen may well consider whether they prefer participation in government by the Fifth Estate to the benefit of all or control of government by labor unions in the interest of labor.

Since most of the troubles that beset mankind have their origin in human nature it would seem worth the while of those who make our laws to study and apply the findings of the biologists and psychologists as to what human nature really is and the springs of its motivation.

Plato called democracy "the best form of bad government." It will be the best form of good government only as it develops the capacity to breed leaders and the faith to trust them. The quality of our children will determine the quality of our democracy. If our laws and *mores* and economic structure continue to discourage breeding from our best strains, if there is to be no adequate recompense for service of the higher types, the time is not far distant when democracy will no longer be safe for the world. If the Fifth Estate were everywhere to be wiped out, as it has been in Russia, the result would be vastly more calamitous than universal war.

Oswald Spengler, in a recent monumental work, forecasts the downfall of Western civilization and would prove his thesis by the history of past cultures. But never in the past has man lived in so compact a world, never has he had such facilities for intercommunication with his fellows, never has he been endowed with such control of natural forces. He has never known himself so well and, above all, never before has he had it in his power to direct so definitely the course of his own development.

Our civilization is certainly imperiled, but there will be no downfall if mankind can be taught to follow the light already before it. As lantern-bearers, it is the clear duty of the Fifth Estate to show the way.

In the past the world has suffered grievously from lack of knowledge; to-day it suffers from its rejection or misapplication. Could the springs of human conduct and the affairs of peoples be regulated only as wisely as we now know how, there would be work and leisure and decent living for all. The criminal, the defective, the feeble-minded would be breeded out, and sane minds in sound bodies breeded in. The loss and suffering from preventable disease and accident would not be tolerated. Higher standards would govern the selection for the public service. Planning would replace *laissez-faire* development, and a rational conservation check the reckless waste of our resources. Production and distribution would attain to levels of efficiency altogether new, and the many injustices now existent in human relations would well-nigh disappear. With the reaction of a freed intelligence on politics, religion, morals, we might hope for a broader tolerance, a better mutual understanding. Wth the recognition of the spirituality of science and the divinity of research and discovery should come larger interests and a new breadth of vision to the average man, and to us all acknowledgment of the steadfast purposive striving shown in the development of the created world, and a reverent appreciation of man's privilege to aid and further this development.

We might reasonably expect ugliness to be replaced by beauty in our cities and small towns and later even in our homes. Government by intelligence for the general good of all should supersede government by special interests, blocs, faddists, and fear of organized minorities and the uninformed crowd. With it all would come relief from the economic pressure which now bears so heavily upon the Fifth Estate that its children, who should be counted among the best assets of the community, are a luxury.

The world needs most a new tolerance, a new understanding, an appreciation of the knowledge now at hand. For these it can look nowhere with such confidence as to the members of the Fifth Estate. Let us, therefore, recognize the obligation we are under. Ours are the duty and privilege of bringing home to every man the wonders, the significance, and the underlying harmony of the world in which we live, to the end that all undertakings may be better ordered, all lives enriched, all spirits fortified.

INTRODUCTORY NOTE

In "Science and Modern Life," Robert Millikan, like Arthur D. Little, called for Americans to use science to eliminate ignorance and corruption and to guarantee the march of progress. He would have resolved the doubts raised by the war by having scientists exercise more influence over societal affairs. Millikan was disturbed by evidence of public antagonism toward science and especially by the call of the British Bishop of Ripon for a ten-year holiday in science. He was convinced that "it is literature and art, much more than science, which have been the prey of those influences through which the chief menace to our civilization comes." The scientific method, Millikan insisted, was the highest expression of reason, and it saved men of science from the pernicious influences flowing from emotionalism. Man's increased ability, gained from science, to consciously make over the physical and biological world fascinated him. Millikan, unlike Russell, did not doubt that the man-made world would be more desirable for humans than the natural world. Millikan, like Little, did not see fault in science or technology, but in a society that failed to fully live by and use them.

Robert A. Millikan (1868–1953) studied physics at Columbia University (receiving a Ph.D. there in 1895) and at the universities of Berlin and Göttingen. Afterward he became a professor at the new and rapidly rising University of Chicago. During World War I, he assisted in mobilizing scientists for war work and was vice-chairman of the National Research Council. In 1920 he was persuaded to accept the directorship of the laboratory of physics at the Throop College of Technology, Pasadena, California. He soon became chairman of the executive council there and played a major part, before his retirement in 1945, in the transformation of Throop into an international center of science and engineering (the name was later changed to California Institute of Technology). In 1923 he received the Nobel Prize for his investigation of the electron and photoelectric effect. In later years, he wrote often for a wide audience on science and society and on science and religion.

Science and Modern Life

ROBERT A. MILLIKAN

I

Last summer it was my lot to be called out of my laboratory to attend in rapid succession (1) a meeting of the Committee on Intellectual Coöperation of the League of Nations at Geneva, a body called into being for the sake of assisting in laying better foundations for international good will and understanding than have heretofore existed; (2) the annual meeting of the British Association for the Advancement of Science at Leeds, one of the most important of the Old World's scientific bodies, whose meetings have marked the milestones of scientific progress; and (3) the International Congress of Physics held at Como and at Rome in commemoration of the hundredth anniversary of the death of Alessandro Volta, the discoverer of current electricity, and thus, in a certain sense, the initiator of this amazing electrical century—suitable errands to inspire reflections on the place of science in modern life. I should like to present them in the form of a few pictures.

As we sped, a thousand persons, across the Atlantic in an oil-burning ship in which even the modern stoker—whose "hard fate" has often been held up as a symbol of the evils of our "mechanical age"—has now a comfortable and an interesting job, for he simply and quietly guides the expenditure of hundreds of thousands of manpower represented in the energy of separated hydrogen and oxygen and carbon atoms rushing eagerly together to fulfill their predetermined destiny, and merely incidentally in so doing sending the ship racing across the Atlantic—in the face of that situation, could I, or could anyone not completely blind to the significance of modern life, fail to reflect somewhat as follows? If Cicero, or Pericles, or any man of any preceding civilization, had been sent on a similar errand, had he had any power at all except the winds, it would have been the man-power furnished to the

The Atlantic Monthly, 141 (1928), 487–496. Reprinted by permission.

triremes by the straining sinews of hundreds of human slaves chained to their oars, slaves to be simply cast aside into the sea, if they weakened or gave out, and then replaced by other slaves! Could any man fail to reflect that our scientific civilization is the first one in history which has not been built on just such human slavery, the first which offers the hope, at least, and a hope already partially realized, of relieving mankind forever from the worst of the physical bondage with which all civilization have heretofore enchained him, whether it be the slavery represented by Millet's man with the hoe—a dumb beastlike broken-backed agricultural drudge—or the slavery at the galley pictured in *Ben Hur*, or the slavery of the pyramid builders referred to in the books of Moses?

Or again, could anyone who stood with me at the base of the column of Trajan, matchless relic and symbol of the unequaled magnificence of Rome, fail to muse first that ancient man in the immensity and daring of his undertakings, in the grandeur of his conceptions, in the beauty and skill of his workmanship, in his whole intellectual equipment, was fully our equal if not our superior? For we shall leave no monuments like his. But could he also fail to reflect that ancient man built these monuments solely through the unlimited control of enforced human labor, while we have not only freed that slave, but have made him the master and director of the giant but insensible Titans of the lower world? We call them now by the unromantic names Coal and Oil. It is our triumph over these that has given to him freedom and opportunity. This is one side of the picture of science and the modern world, a side that can be presented with a thousand variants, but all having the same inspiring significance.

Another picture. In a comfortable English home out in the country in North England a small group is seated, sipping after-dinner coffee, enjoying conversation, and interrupting it now and then to listen to something particularly fine that is coming in over the radio. The technique of the reproduction is superb, but no more so than that with which we are familiar in our American homes, for the whole broadcasting idea, as well as the main part of its technical development, is American in its origins. But the programme that is on the air in England is incomparably superior to anything to be heard here, for the English Government has taken over completely the control of the radio. It collects from each owner of a receiving set twelve shillings a year, and then, with the large funds thus obtained, —for there are many radio fans in England as in America,— it provides the radio-land public of England with the largest return in education and in entertainment for eight mills a night ever provided, I suspect, anywhere in the history of the world. For it employs only high-class speakers, musicians, and entertainers of all sorts, so that the whole British nation is now being given educational advantages of the finest possible sort through the radio, at less than a cent a family a night, collected only from those who wish to take advantage of them.

Nor is it merely the subject matter of the radio programmes that is commendable. The value of giving the whole British public the opportunity to hear the English language used, in intonations and otherwise, as cultured people are wont to use it, is altogether inestimable. And, sitting there in the North of England, we had but to turn the dial to the wave length used by Berlin and we heard an equally authoritative use of the German language, and I envisaged a whole population, or as many of it as wished, learning a new language, easily and correctly, instead of through the stupidities of grammar, as we now go at it. What a stimulant to the imagination! What possibilities are here, only just beginning to be realized, for public education, for the enrichment of the life of the country dweller, as well as the city resident, solely because of such an influence as this of modern physics upon modern life.

II

Now turn to another picture which presents the other side of the story. Sir Arthur Keith, the foremost British anthropologist, is now the president of the British Association. The Leeds meeting represented the fiftieth anniversary of the meeting at which Darwin's then new theory of evolution had been first vigorously debated. Sir Arthur took last summer, as the subject of his presidential address, "Darwin's Theory of Man's Descent As It Stands To-day." He showed that fifty years of fossil study had given extraordinary confirmation to the general outline of the evolutionary conception, had placed it, indeed, upon well-nigh impregnable foundations.

The following Sunday the Bishop of Ripon preached upon

science and modern life. He thought we were gaining new scientific knowledge, and acquiring control of stupendous new forces, faster than we were developing our abilities to control ourselves, faster than we were exhibiting capacity to be entrusted with these new forces, and hence he suggested that science as a whole take a ten-year holiday.

When, the next day, the newspaper men, who had had as good a story out of the whole incident as our newspapers got out of the Scopes trial, pressed the Bishop to define more sharply just what he meant by a ten-year scientific holiday, he was reported to have said that he thought the workers in medicine and in public health ought not to stop, since then the germs of disease might steal a march on us, and avoidable suffering be thereby caused. He had had in mind, rather, a vacation for physics and chemistry and the parts of biology not associated with the improvement of health and the alleviation of suffering.

The Bishop's explanation is of value as throwing an illuminating side light upon the sort of emotionalism and misunderstanding that is represented in much of the present public antagonism to our scientific progress. The question which the Bishop raises is proper enough, but the conclusion is altogether incorrect. For, first, physics and chemistry cannot take a holiday without turning off the power on all the other sciences that depend upon them, for biological science is at bottom only one of the applications of physics and chemistry; and, second, physics, chemistry, and genetics are, in fact, the great, constructive sciences which alone stand between mankind and its dire fate foreseen by Malthus. The palliative sciences, such as the Bishop mentioned, are indeed worthy of support, but without the fundamental sciences they only hasten and make inevitable the horrors of that day.

The incident is presented because it is illustrative of a widespread attitude as to the danger of flooding the world with too much knowledge. The fear of knowledge is quite as old as the Garden of Eden. Prometheus was chained to a rock and had his liver torn out by a vulture because he had dared to steal knowledge from the gods and bring it down to men. The story of Faust, which permeates literature up to within a hundred years, is evidence of the widespread, age-long belief in the liaison between the man of knowledge and the powers of darkness. It will persist so long as superstition, as distinct from reverence, lasts.

But there is a real question, not to be thus easily disposed of, which the Bishop's sermon puts before the man of science. It is this. "Am I myself a broadly enough educated man to distinguish, when I am engaged in the work of reconstruction, between the truth of the past and the error of the past, and not to pull them both down together? Am I sufficiently familiar with what the past has learned, and what it therefore actually has to teach, and am I enough of a statesman not to remove any brick from the structure of man's progress until I see how to replace it by a better one?" I am sorry to be obliged to admit that some of us scientists will have to answer that question in the negative. Such justification as there may be for the public's distrust of science is due chiefly to the misrepresentation of science by some of its uneducated devotees. For men without any real understanding are of course to be found in all the walks of life.

This problem, however, is not at all peculiar to science. In fact, the most wantonly destructive forces in modern life, and the most sordidly commercial, are not in general found in the field of science, nor having anything to do with it. It is literature and art, much more than science, which have been the prey of those influences through which the chief menace to our civilization comes. After the law of gravitation, or the principle of conservation of energy, has been once discovered and established, physics understands quite well that its future progress must be made in conformity with these laws, at least that Einstein must include Newton, and it succeeds fairly well in keeping its levitators and its inventors of devices for realizing perpetual motion under suitable detention, or restraint, somewhere. But society has as yet developed no protection against its perpetual-motion cranks—the devotees of the new, regardless of the true—in the fields of literature and art, and that despite the fact that sculpture has had its Phidias and literature its Shakespeare just as truly as physics has had its Newton or biology its Pasteur.

I grant that in literature and art, and in nonscientific fields generally, it is more difficult than in science to know what has been found to be truth and what error, that in many cases we do not yet know; nevertheless there are even here certain broad lines of established truth recognized by thoughtful people everywhere. For example, the race long ago learned that unbridled license in the individual is incompatible with social progress, that civilization, which is orderly group life, will perish and the race go back to the jungle unless the sense of social responsibility can be kept universally alive. And yet today literature is infested here and there with unbridled license, with emotional, destructive, oversexed, neurotic influences, the product of men who either are incompetent to think anything through to its consequences or else belong to that not inconsiderable group who protest that they are not in the least interested in social consequences anyway, men who, in their own words, are merely desirous of "expressing themselves." Such men are, in fact, nothing but the perpetual-motion cranks of literature and of art. It is from this direction, not from the direction of science, that the chief menaces to our civilization are now coming.

But, despite this situation, I should hesitate to suggest that all writers and all artists be given a holiday. This is an age of specialization, and properly so, and some evils from our specialization are to be expected. Our job is to minimize them and to find counterirritants for them. I am not altogether discouraged even when I find a humanist of the better sort who is only half-educated.

Let this incident illustrate. Not long ago I heard a certain British literary man of magnificent craftsmanship and fine influence in his own field declare that he saw no values in our modern "mechanical age." Further, this same man recently visited a plant where the very foundations of our modern civilization are being laid. A ton of earth lies underneath a mountain. Scattered through that ton in infinitesimal grains is just two dollars' worth of copper. That ton of earth is being dug out of its resting place, transported to the mill miles away, the infinitesimal particles of copper miraculously picked out by invisible chemical forces, then deposited in great sheets by the equally invisible physical forces of the electric current, then shipped three thousand miles and again refined, then drawn into wires to transport the formerly wasted energy of a waterfall—and all these operations from the buried ton of Arizona dirt to refined copper in New York done at a cost of less than two dollars, for there was no more value there.

This amazing achievement not only did not interest this humanist, but he complained about disfiguring the desert by electrical transmission lines. Unbelievable blindness—a soul without a spark of imagination, else it would have seen the hundred thousand powerful, prancing horses which are speeding along each of those wires, transforming the desert into a garden, making it possible for him and his kind to live and work without standing on the bowed backs of human slaves as his prototype has always done in ages past. Seen in this rôle, that humanist was neither humanist nor philosopher, for he was not really interested in humanity. In this picture it is the scientist who is the real humanist. Nevertheless, the Bishop of Ripon was right enough in distrusting the wisdom, and sometimes even the morality, of individual scientists, and of individual humanists, too. But the remedy is certainly not to "give science a holiday." That is both impossible and foolish. It is rather to reconstruct and extend our educational processes so as to make broader-gauge and better-educated scientists and humanists alike. There is no other remedy.

III

But, says someone, these pictures so far deal only with the superficial aspects of life. What has science to say to him whose soul is hungry, to him who cries, "Man shall not live by bread alone"? Has it anything more than a dry crust to offer him? The response is instant and unambiguous. Within the past half century, as a direct result of the findings of modern science, there has developed an evolutionary philosophy—an evolutionary religion, too, if you will—which has given a new emotional basis to life, the most inspiring and the most forward-looking that the world has thus far seen. For, first, the findings of physics, chemistry, and astronomy have within twenty-five years brought to light a universe of extraordinary and unexpected orderliness, and of the wondrous beauty and harmony that go with order. It is the same story whether one looks out upon the island universes brought to light by modern astronomy, and located definitely, some of them, a million light years away, or whether he looks down into the molecular world of chemistry, or through it to the electronic world of physics, or peers even inside the unbelievably small nucleus of the atoms. Also, in the organic world, the sciences of geology, palæontology, and biology have revealed, still more wonderfully, an orderly development from lower up to higher forms, from smaller up to larger capacities—a development which can be definitely seen to have been going on for millions upon millions of years and which therefore gives promise of going on for ages yet to be.

A fire-mist and a planet,
A crystal and a cell,
A jellyfish and a saurian,
And caves where the cavemen dwell;
Then a sense of law and beauty,
And a face turned from the clod—
Some call it Evolution,
And others call it God.

That sort of sentiment is the gift of modern science to the world.

And there is one further finding of modern science which has a tremendous inspirational appeal. It is the discovery of the vital part which we ourselves are playing in this evolutionary process. For man himself has within two hundred years discovered new forces with the aid of which he is now consciously and very rapidly making over both his physical and his biological environment. The Volta Centenary, a symbol of our electrical age, was representative of the one, the stamping out of yellow fever is an illustration of the other. And if the biologist is right that the biological evolution of the human organism is going on so slowly that man himself is not now endowed with capacities appreciably different from those which he brought with him into the period of recorded history, then since, within this period, the forward strides that he has made in his control over his environment, in the development of his civilization, have been stupendous and unquestionable, it follows that this progress has been due, not to the betterment of his stock, but rather primarily to the passing on of the accumulated knowledge of the race to the generations following after. The great instruments of progress for mankind are then research—the discovery of new knowledge—and education—

the passing on of the store of accumulated wisdom to our followers. This puts the immediate destinies of the race or of our section of the race, or of our section of our country, largely in our own hands. This spirit and this conviction are the gift of modern science to the world. Is it, then, too much to say that modern science has remade philosophy and revivified religion?

IV

The next picture brings into the foreground what I regard as the most important contribution of science to modern life. The scene is laid in Geneva; the occasion, a meeting of the Council of the League of Nations. The speaker is Nansen, the tall, white-haired, rugged-faced, heavily moustached Norwegian explorer, now directing the tamed and controlled energies of his fierce viking blood to trying to find a solution to the tragic situation of the Armenians, a situation to which heretofore there has been no solution except extermination. After four years of effort he brings in a discouraging report, and thinks the League of Nations must write down 'the record of its first failure.' He requests the Council to strike the Armenian matter from its programme, promising, however, to keep at it himself and to try through other agencies to find a solution. Then Briand of France speaks. Quietly he begs Nansen not yet to despair of the League's assistance. He is sure some solution can be found, and promises that his country, in financial straits though it be, will not be lacking in lending its assistance. The representatives of other nations follow in similar vein, the problem is retained on the Council's programme, and the conviction is at least fortified that with the right kind of attack a solution may yet be found to an age-old difficulty—that extermination is not the only answer to race rivalry.

With the right sort of attack! What is it that the League of Nations as a whole is trying to do? It is trying for the first time in human history to use the objective mode of approach to international difficulties, in the conviction that there is some better solution than the arbitrament of war. But whence has come that conviction? Without the growth of modern science it certainly would have been slower in coming. Perhaps it would not have come at all. In the days of the jungle, war *was* probably the best solution—at least it was the only solution. It was Nature's way of enabling the fittest to survive, and we are not so far past the days of the jungle yet. Within fifty years as great an historian as Eduard Meyer and as great a humanist as John Ruskin have lauded war as the finest developer of a people. But it has been quite recently demonstrated that war is no longer, in general, the best way to enable the fittest to survive.

The Great War profited no one. It injured all the main participants. Modern science has created a new world in which the old rules no longer work.

> New occasions teach new duties, time makes
> ancient good uncouth;
> They must upward still and onward, who would
> keep abreast of truth.

Alfred Nobel was perhaps not far from right when he thought that he had taken the main step to the abolition of war by the invention of nitroglycerine. He has, I suspect, exerted a larger influence in that direction than have all the sentimental pacificist organizations that have ever existed. For sentimental pacificism is, after all, but a return to the method of the jungle. It is in the jungle that emotionalism alone determines conduct, and wherever that is true no other than the law of the jungle is possible. For the emotion of hate is sure sooner or later to follow on the emotion of love, and then there is a spring for the throat. It is altogether obvious that the only quality which really distinguishes man from the brutes is his reason. You may call that an unsafe guide, but he has absolutely no other unless he is to turn his face back toward the jungle from which he has come. There is no sort of alternative except to set up in international matters, precisely as we have already done in intercommunity and interstate affairs, some sort of organization for making studies by the objective method of international difficulties and finding other solutions.

But what exactly do I mean by the objective method? Somebody has said that "what we call the process of reasoning is merely the process of rearranging one's prejudices," and we admit the truth of this assertion when we say, as we so often do, "Oh yes, I understand that is the excuse, but what is, after all, the reason?" Indeed, there is no question that a large part of what we call reasoning is in fact simply the rearranging of prejudices. In so far, for example, as we are Republicans, or Democrats, or Presbyterians, or Catholics, or Mohammedans, or prohibitionists, because our fathers bore those brands,—and many of us will be admitted by our acquaintances, at least, to have no other real grounds for our labels,—our so-called reasonings on these subjects certainly consist in nothing more than the rearrangement of our prejudices. The lawyer who takes a case first, and develops his argument later, is obviously only rearranging his preconceptions.

If, however, one wishes to obtain a clear idea of what the objective method is, he has only to become acquainted with the way in which all problems are attacked in the analytical sciences. In physics, for example, the procedure in problem solving is always first to collect the facts—namely, to make the observations with complete honesty and complete disregard of all theories and all presuppositions, and then to analyze the data to see what conclusions follow necessarily from them, or what interpretations are consistent with them. This method, while not confined at all to the physical sciences, is nevertheless commonly known as the scientific method in recognition of the fact that it has had its fullest development and its most conspicuous use in the sciences. Indeed, I regard the development and spread of this method as the most important contribution of science to life, for it represents the only hope of the race of ultimately getting out of the jungle. The method can in no way be acquired and understood so well as by the study of the analytical sciences, and hence an education which has left out these sciences has, in my judgment, lost the most vital element in all education.

Nor is that merely the individual judgment of a prejudiced scientist, as the following quotation from one of our most prominent humanists, a member of the faculty of Harvard University, shows:—

> It is the glory of pure science and of mathematics that these subjects train men in orderly and objective thinking as no other subjects can. Here are fields of study in which loose or crooked thought leads inevitably to demonstrable error, to error which cannot be glossed over or concealed. Here are branches of knowledge in which there is no confusion between right and wrong, between post hocs and propter hocs, between the mere coincidences and the consequences of a cause. When you have

finished with a problem in any of the exact sciences you are either right or wrong, and you know it. That is why we call them exact sciences, to distinguish them from philosophy, sociology, economics, and the other social sciences, in which the difference between truth and error is still, in most cases, a matter of individual opinion. Many years ago physics was known as "natural philosophy"; it was merely a body of speculative ideas concerning the mechanics of nature. It became an exact science by developing an inductive methodology, which makes all the difference between science and guesswork.

Some years ago, in the Harvard Law School, we thought it worth while to inquire into the educational antecedents of the student body, with a view to ascertaining whether there was any relation between success in the study of law and the previous collegiate training of these young men. In the Harvard Law School there are more than a thousand students, all of them college graduates, drawn from every section of the country. Nearly all of them have specialized, during their undergraduate years, in some single subject or group of subjects—languages, history, science, philosophy, economics, mathematics, and so on. Off-hand one would probably say that the young man who had devoted most of his attention to the study of history, government, and economics while in college would be gaining the best preparation for the study of law—for these are the subjects which in their content come nearest to the law; but that is not what we found. On the contrary the results of this inquiry showed that the young men who had specialized in ancient languages, in the exact sciences, and especially in mathematics, were on the whole better equipped for the study of law, and were making higher rank in it, than were those who had devoted their energies to subjects more closely akin.

But can education, even in the sciences, do the work fast enough to prevent the catastrophe feared by the Bishop of Ripon? Can we learn to control our emotions and impulses and our new-found powers, to take the long view, and to do the rational thing instead of the emotional, or the vicious, thing with the enormous forces given to us by science? Can we alter human nature?

Perhaps the following is a partial answer: twenty-five years ago if anyone had asked you or me or any body of men, however intelligent, whether human nature could be so altered in a reasonable time as to make it safe to entrust practically every grown man in California and part of the women and children with a thiry-horsepower locomotive which they might drive at will through the crowded cities, and race at express-train speed over the country roads of California, the answer would certainly have been a decided negative. Nobody on earth, I suspect, would have thought such a result possible. And yet that is precisely what has happened. It is true we have accidents, too many of them by far. There is still much to improve, and yet the risk is so small that we never think of it when we enter an automobile. I marvel at the success of it every time I drive in city streets. I glory in it when I see the new race of men the taxi business has created in a city like London. Contrast the clear-eyed, sober, skillful, intelligent-looking London taxi-driver of to-day with the red-nosed wreck of a human being who used to be the London cabby of a quarter century ago, and see what responsibility and power do in altering human nature.

Also the picture which modern science has unfolded of the age-long history of the biological organism is one in which it is seen adapting itself with marvelous success to changes in external conditions. That we, too, at our end of this evolutionary scale, have inherited this adaptability was one of the most striking lessons of the late war, in which we settled down to the endurance of what we thought intolerable conditions with amazing rapidity.

V

If, then, there be any notes of optimism in modern life, one of them is certainly the note played by modern science. If there be any escape from Malthusianism,—the world's greatest problem,—science alone can provide it. It is clearly the development of science and its application to modern life that have made possible the support in Great Britain to-day of forty million people, when a hundred and fifty years ago Benjamin Franklin called England overpopulated with eight millions, and when Robert Fulton a little later, in a prophetic mood, saw England holding sometime "a population of ten million souls." Perhaps this population has to-day gone too far, but the check is being applied. In both England and Sweden to-day the birth rate is less than it is in France. With the creative power of physical science, and the application of intelligence to the findings of biological science, even this problem of population can be faced with a good measure of hope. An international union for its continuous study was formed last summer at Geneva. That is the objective way to begin to attack it.

Finally, can science save our civilization from the fate that has befallen its predecessors, the Sumerian, the Egyptian, the Greek, the Roman, and the others that have risen and then declined? Are the Keyserlings, the Spenglers, and the other prophets of decay and death to be regarded as real prophets?

The answer is, of course, a secret of the gods, but that these "prophets" are "multiplying words without knowledge" it is easy to show. For our modern civilization rests upon an altogether new sort of foundation. These older civilizations have rested upon the discovery of new fields of knowledge or of art—fields which the discoverers have indeed cultivated with such extraordinary skill that they have been able to reach a state of perfection in them that succeeding generations have often been unable to excel. Witness the sepulchral art of the Egyptians, and the perfection of such principles of architecture as they knew how to use; witness the sculpture, the painting, the æsthetic and the purely intellectual life of the Greeks—an accomplishment so great as to inspire an outstanding modern artist to say that there has been no new principle discovered in either sculpture or painting since the age of Pericles; witness the principles of government and of social order discovered by the Romans, or the arch in architecture, altogether Roman, but reaching perhaps its perfection in the Romanesque and the Gothic of a few centuries later; witness the discovery of the principles of music in central and southern Europe in the Middle Ages, and the perfection that art attained within two or three centuries. And let us remember, too, that humanity for all time is the inheritor of these achievements. This is the truth of the past which it is our opportunity and our duty to pass on to our children.

But our modern world is distinctive not for the discovery of new modes of expression or new fields of knowledge, though it has opened up enough of these, but for the discovery of the very idea of progress, for the discovery of the method by which progress comes about, and for inspiring the world with confidence in the values of that method. So long as the world can be kept thus inspired, it is difficult to see how a relapse to another dark age can take place.

Even if the biological evolution of the human race should not continue,—though why should what has been going on for millions of years have come to an end just now?—yet the process by which progress has been made within historic times can scarcely fail to be continuously operative. This process is the discovery of new knowledge by each generation and the transmission to the following generation of the accumulated accomplishment of the past—the discovery of new truth and the passing on of old truth.

The importance of both elements in this process has not been realized in the past, and dark ages have come. But the means for the spread of knowledge, for its preservation and transmission, the facilities for universal education and inspiration, the time for leisure and the opportunity for thought for everybody—all these have been so extended by modern science, and are capable of such further extension, that no prophecy of decline can possibly have any scientific foundation. Even arguing solely by the method of extrapolation from the past, modern science has shown that the ups and downs on the curve of history are superposed upon a curve whose general trend is upward, and it has therefore brought forth a certain amount of justification for the faith that it will continue to be upward. In the last analysis, humanity has but one supreme problem, the problem of kindling the torch of enlightened creative effort, here and there and everywhere, and of passing on for the enrichment of the lives of future generations the truth already discovered—in two words, the problem of research and of education.

INTRODUCTORY NOTE

Arthur H. Compton, Nobel Laureate physicist, in "Oxford and Chicago: A Contrast," responded to the growing ambivalence, even antipathy, toward science and technology. He found the antipathy especially strong at Oxford University, where British scholars determinedly guarded their traditional life style and philosophy, which were grounded in literature and the arts, against the influence of science and technology. Although Compton respected the traditional values they were attempting to sustain, he saw a more positive approach taking shape in Chicago, which to him represented the essence of twentieth-century America. It accepted science and technology and embraced the new world being shaped by them. He foresaw Chicago helping to fashion a society that would use technology and the scientific method as means of fulfilling not only material needs but spiritual aspirations as well. Realizing that some outcries against technology stemmed from those who feared that men were being forced into the Procrustean bed of machine-dominated systems, Compton prophesied that science and technology in the new Chicago would allow individuals to express their unique personalities. He believed, like Little and Millikan, that the philosophers and the leaders of the transition to the new world would be scientists and university scholars.

Arthur Holley Compton (1892–1962) received a Ph.D. degree in physics from Princeton University in 1916. After a year of research at the Cavendish Laboratory of Cambridge University, he became a professor at Washington University, St. Louis, and later at the University of Chicago. In 1927 he received the Nobel prize for physics. During World War II, he was director of the metallurgical project at the University of Chicago that contributed substantially to the development of the atomic bomb. Throughout his life he was interested and active in cultural and religious affairs, chairing the National Conference of Christians and Jews and helping to plan UNESCO.

Oxford and Chicago: A Contrast

ARTHUR H. COMPTON

An eminent scientist views the classic traditions and way of life at Oxford as contrasted with the technological foundation of Chicago society.

An English friend, after a brief visit in Chicago, remarked that he found its life unbearably superficial. The total lack of tradition, the complete absorption of effort in the machinery of life! How could people endure an existence so empty of human values?

A week earlier my wife and I had dined in Oxford with a leader of British education. The question arose regarding the place of science in life. It had proved itself as a method for arriving at reliable knowledge, and for supplying us with certain necessaries of life—but the inhumanness of it all! The discipline of thought inculcated by science was already embodied in Plato's mathematics. The predominant trend of modern science is toward replacing the human interests present in literature, art and music with technological developments in which the human factor becomes less and less significant. The most fundamental bases of morality and religion have been ruthlessly shaken, with the implication that their value is insignificant. In place of a quiet ramble over the varied English countryside, we have a powerful motor car tearing down the broad hard highway. In Oxford, fortunately, there survives the tradition of emphasis on human values.

In Oxford, science and technology form a dark cloud, threatening the very existence of the traditional mode of life which has proved its value through the ages. In Chicago, life is based upon technology and science, which have become an accepted part of life as fundamental as agriculture, and no more to be feared.

Scribner's Magazine, 99 (1936), 355–357. Reprinted by permission.

With but brief history, Chicago shows a spirit of youthful enthusiasm. The powerful methods of technology are used to build what it considers a beautiful type of architecture adapted to modern needs. Its University devoted to the advancement of knowledge in the spirit of science has won its spurs. Its Grand Opera, though financially unfortunate, marked a notable effort to enrich art by the gifts of technology. Though unquestionably more crude in these early stages than the highly refined culture developed through the ages upon the classical background of Greece, no one who has lived in the Midwest of the United States can doubt that it has a spirit which goes far beyond the superficial machinery of life as observed by the casual visitor. Men and women with interests as high as heaven and as deep as the heart of man arise in all communities. I find them among my friends in Chicago as well as among those in Oxford. But their backgrounds are widely different. For the Oxonian the lore of history and literature is the starting point with which an effort is being made to reconcile the facts of science and the changing society based on technology. The Chicagoan postulates the world of science and gropes for a more complete understanding of life using ancient tradition only as an occasional guide.

I have chosen Chicago and Oxford because they represent opposite extremes in the development of Western civilization. Oxford with her scorn of central heating, but her mellow old buildings and beautiful gardens, and her gentlemen of culture and leisure, shows a very different sense of values from Chicago with her innocence of tradition, but her eagerness to try experiments in architecture or education and her rating of citizens according to their accomplishment for the common weal. Chicago, as the metropolis of the Middle West, represents urban Americanism in its purest form, remote from the foreign influences that enrich life near the borders of the country. In a new community new ideas are readily absorbed, and newness becomes a virtue. Oxford, more than any other city, represents to my mind the legacy of the ages of Western culture, and the studied effort to incorporate the most valuable parts of this legacy in the lives of the young men who are to guide the development of the growing world. Thus I am taking Chicago to represent the extreme, and in certain respects the best development of American culture, just as Oxford represents the fine traditions of Europe. They are approaching by different routes the more satisfactory adjustment of life to a world in which man, using the tools of science, is empowered to shape his world as he will.

In India it is possible for Gandhi to persuade his followers that the works of technology are harmful, and should be kept out of their lives. The Oxford don may decry the dehumanizing influence of science, prefer a pen (though a fountain pen) to a typewriter, and walk rather than ride behind a petrol motor. The Chicagoan however cannot question the value of science and technology. They are his very life. Railroads, automobiles, airplanes; agricultural machinery, mechanized meat packing, pasteurized milk; telephones, radios, movies; electric power—if these were gone he could not live. One might as well ask the farmer whether there is any value in planting crops. These things are the basis of existence. The rapid growth of the community gives evidence that many find life in such a world worth while.

None but a hasty consideration can result in a judgment that the achievements of science and technology are of only passing interest. With the present world-wide spread of knowledge it is almost unthinkable that those technological developments which have been found of fundamental human value should be lost. Among such basic crafts we must include modern methods of transportation, communication, heating and lighting, and sanitation. These by themselves would make necessary continued knowledge of the fundamental sciences. It was possible during the Greek and Arabic periods for men to lose their interest in science because it had not proved its practical worth, and in philosophical thought its incomplete form led to insoluble difficulties. I can envisage nothing short of a world-wide catastrophe almost completely destroying mankind which could now do more than make a temporary pause in man's interest in science and technology. If this view is correct, the "mechanized world" which some already feel so oppressive is destined to be the world of the future. Blessed are those who can find in such a world the outlets which their spirits require.

It would be false to conclude that the major interests of mankind will dwell more and more upon technology; rather the reverse. When men first found agriculture profitable, it became the great center of interest, for it was the new basis of civilization. In time, however, agriculture became so much a matter of routine that it ceased to hold this preëminent position. As a result of agricultural development, man's world has been completely changed. Who has not sighed for the lonely peace of the primeval forest or fails to be inspired by the vast expanse of desert sands or the untouched rugged mountains? Yet we no longer live in these surroundings. Our fields are cultivated; our trees are for the most part planted. In place of the primitive beauties of nature, we shape nature to our desires in flower beds and hedgerows. Who would say that the Japanese does not find as great satisfaction in his cultivated garden as the American Indian in the forests through which he used to roam? Likewise we find ourselves developing new tastes for beauty in our machines. I confess a thrill of pleasure from the lines and performance of a well-made automobile quite comparable with that from watching a deer spring lightly through the forest. The fact that the car was built to fit my needs gives me that pride of accomplishment of a job well done which adds to my appreciation. Similarly, with new forms of illuminants, a new art of home lighting arises. New technics lead to new possibilities of expression, and man becomes more of a creator. Gradually we may anticipate less attention being given to the technical methods and their scientific basis, and more emphasis on those refinements which enable us to express our personalities.

Science and technology are thus placing in our hands new and more mighty powers. Inevitably there comes a change in our mode of life comparable with that which accompanied the development of growing grain as a source of food. With increased population, changed social habits, and new codes of morals, human life is being formed anew.

In this new world where we depend more and more upon science and its consequences, it becomes education's primary responsibility to help men find a satisfying way of life in their new surroundings. That spiritual needs exist and are not being satisfied becomes evident from the multiplication of faddist cults, from the frequent outcries against science as the despoiler of religion, or against technology as moulding us into machines. It is not only in the Oxfords that such needs are felt. Men of spiritual aspirations live also in our Chicagos.

Fortunately, the human species is extraordinarily adaptable. Under changed conditions evolutionary changes proceed at an accelerated rate. In the present case this means social evolution. If men are to reach a satisfying life as masters of machines, it cannot be by fighting against a "brave new world" that is thrust upon them, but by finding themselves a part of this world and adapting their lives to it. Those on whom this world merely thrusts itself with irritating frequency may retreat behind walls of tradition and consider academically what adjustments should be made. Those, however, who are a part of the mechanized world must adapt themselves to it or perish. Clearly it is toward the mechanized communities that we must look to find the most rapid adaptation of our thought and customs.

For this reason, as I see it, the educational institutions in our Chicagos are those toward which men must look for leadership in meeting the spiritual as well as the more mundane problems of life in the changing world. Here are communities of scholars schooled in the scientific method. In Oxford, with its "classics" and "modern greats," science plays a minor rôle. In Chicago, three of the four main divisions of the University are sciences: physical, biological, and social; and even in the humanities the scientific approach is close to the surface. At the same time, with four affiliated theological schools, a great Oriental Institute for studying man's growth through the ages, and extensive development in art, we find a conscious effort to find how those things of tested worth can be incorporated into our modern society. To a greater or less extent any of the large universities of the Middle West presents a similar picture. Nor is it confined to our Midwest. If the severity of the demands of mechanized life are not felt quite so strongly along the Atlantic coast, the great educational institutions there located have the advantage of a closer contact with Europe, which must serve as a most valuable guide. Using the powerful methods of science, and sharing the life of the mechanized community, our educational leaders are making every effort to find and develop the true values present in the new world. This search is naturally most urgent and the changes most rapid where science and technology are most essential to life.

In the senior fellows' common room of one of the Oxford colleges, I was seated between two classics dons. Trying to find what was included in a well-rounded course in classics, I inquired whether their students looked into the question of the influence of Oriental mysticism in changing Greek thought after Alexander's eastern conquests. "That is the kind of subject a research student might investigate as a specialty," was the reply, "but it would be out of place in our classics courses. For you see, we have carefully worked over the ancient writings, and we mean by the classics those writings which have been found of greatest value. It would waste the time of our students to divert their attention toward other works." Tradition, yesterday, today, forever; conservator of ancient values, without a glimpse of changing conditions!

One of the finest products of civilization is the well-organized life found in our Oxfords, mellowed by the appreciation of human values as revealed by the thought of centuries of scholars and as cultivated by many generations of men of leisure. The legacy of past ages is used in such communities to enrich the life of today. In many ways the life of Oxford has a completeness that is not approached in our more modern and supposedly more sophisticated cities. Our world could ill afford to lose the leaven supplied by these centers of culture.

Yet in the industrial communities where technological civilization is at its height, the life of Oxford cannot be lived. It is only by setting itself apart from the world both physically and intellectually that this charming community has retained its traditional values. What the world needs is the building of these values or their modern equivalent into the structure of twentieth-century life. It is this objective which has been seen perhaps more clearly by the leading universities of our central United States than by any other important educational group. Our Chicagos must look for spiritual leadership to universities whose men are themselves a part of the machine age, but who have learned to take into their own lives the legacies transmitted from old cultures.

Nor do I believe this hope of leadership is vain. Almost unconsciously perhaps a philosophy of life appropriate to modern needs is formulating itself on the campuses of our American universities. Science and the freedom to study all subjects by the scientific method are assumed without question. We thus have more effective tools for the interpretation of life. The astronomer opens to us a new and mightier universe. The physicist finds unsuspected limits to our knowledge which promise to simplify the problems of the philosophers. The chemist applies his powerful methods to the problems of life, while the professor of physiology makes a religion of his science. The sociologist studies the development of human life in our mechanized community, while the theologian aspires to express his ideas so as to be acceptable to his scientific colleagues. Over the round-table at the club these various points of view are matched to throw into clearer perspective the human problems of the day. Every man in the group is engaged in reformulating his own world view. Perhaps such thinking is unorthodox; but in a very real sense I believe it is the most vital and concentrated "religious" thinking that has been done since the Reformation.

Those who have watched developments since the war cannot fail to be impressed by the way in which the leadership in almost all branches of science has passed to the United States. Germany has resigned her one-time preëminence. France, Italy, Sweden, and other European countries continue to do good work, but on a relatively small scale. Per capita I suppose England at least matches our efforts; but the many research institutions in our country easily outstrip her in the value of their total output. Russia's intense interest in technology and science has not yet carried her to the stage of great achievement. The mantle of leadership in science rests upon our shoulders.

Where then if not toward American education must a scientifically minded world look for spiritual leadership? Education even more than technology is passing through a formative period. Though technology has yet far to go before its task is complete, the direction it will follow, of making the world a social and economic unit in which the physical needs of man are satisfied in a more abundant manner, seems evident. The moulding of our lives to make the best of such a world has however hardly begun. Science is giving us clearer vision with which to see and appreciate the world of which we are a part. We can begin to see more clearly outlined the great plan according to which that world is developing. The goal we do not know. But in some mysterious way, we find ourselves able to take part in shaping this world. More and more the responsibility for the life upon this planet, and even for our own evolution, is being transferred to us. Our great need is a truer understanding of life's values, so we may wisely choose the direction of our own development.

A thorough examination of life's great values is thus urgently presented as the primary task of education. Old formulas derived at our Oxfords cannot be applied directly in our Chicagos, though they may be most helpful in pointing the way. Here is one of history's greatest opportunities for social pioneering—the discovery and development of human objectives which will enable men to use to their full value the new powers with which science has gifted them. American universities are the place, and now is the time for this great human problem to be solved.

INTRODUCTORY NOTE

Howard Scott, leader of the Technocracy movement that briefly attracted broad public attention during the depression-ridden thirties, attempted to explain Technocracy to the public in "Technocracy Speaks." Because of unorthodox terminology, his argument is not readily comprehensible, but his thesis is simple: he called for a reorganization of society, especially its economic structure, so that technological potential could be fulfilled. While Scott avoided making explicit a radical demand for government by engineering and scientific experts, his program was, in fact, far more radical than the "scientism" of Little and Millikan. They wanted society changed by bringing scientific experts into society's existing decision-making positions; Scott wanted a drastically reordered economy and society based on the availability of energy and its utilization to produce goods and services. His was the relentless logic of engineers and efficiency experts. At the time of the Great Depression, when factories were closed and people were idle, cold, and hungry, Scott's drastic rationalism had a natural appeal.

The history of Scott and Technocracy in the early thirties was one of a meteoric rise followed by an almost as meteoric fall with the coming of Franklin D. Roosevelt's New Deal reforms. These were a less radical response to the diseconomies fixed upon by Scott and his associates. Then critics pointed out factual errors in the Technocrats' master plan for the reorganization of the economy—the "Energy Survey of North America"—and questioned Scott's engineering credentials. Finally, leading engineers, those who were to implement the plan, dissociated themselves from the Technocracy program.

Howard Scott in 1932 was described sympathetically as a forty-two-year-old Virginian with a degree of Doctor of Engineering from the University of Berlin. He had reportedly been a munitions expert in Canada during World War I, a research scientist in the dye industry in Germany earlier, and a leading technician on the Muscle Shoals project in the United States after the war. Critics said none of this was true: In the case of Muscle Shoals, Scott had been the foreman of a cement-pouring gang. He had been a partner in a floor-wax manufacturing firm, it was contended, and had become better known as a Greenwich Village figure eager to advocate theories of social change.

Technocracy Speaks

HOWARD SCOTT

The very favorable editorial comment on Technocracy presented by Mr. Orage in a recent issue of the *New English Weekly* has prompted the editor of THE LIVING AGE to request a statement from Technocracy itself. I should like first to make two corrections in matters of fact. First, the per capita consumption of energy should be respectively 2,000 and 150,000 kilogramme calories *per day,* not per annum as Mr. Orage stated. Secondly, the cigarette production is given as at the rate "of five or six hundred per hour per man," instead of per minute per man.

The following communication is less a reply to Mr. Orage's comments than a statement of the aims and methods of the research body known as Technocracy. Such a statement would seem appropriate after the very considerable publicity given to the specific findings of Technocracy in connection with its Energy Survey of North America.

I

Technocracy is a research organization, founded in 1920, composed of scientists, technologists, physicists, and biochemists. It was organized to collect and collate data on the physical functioning of the social mechanism on the North American continent, and to portray the relationship of this continent and the magnitude of its operations in quantitative comparison with other continental areas of the world. Its methods are the result of a synthetic integration of the physical sciences that pertain to the determination of all functional sequences of social phenomena.

Technocracy makes one basic postulate: that the phenomena involved in the functional operation of a social mech-

The Living Age, 343 (December 1932), 297–303.

anism are metrical. It defines science as "the methodology of the determination of the most probable." Technocracy therefore assumes from its postulate that there already exist fundamental and arbitrary units which, in conjunction with derived units, can be extended to form a new and basic method for the quantitative analysis and determination of the next most probable state of any social mechanism. Technocracy further states that, as all organic and inorganic mechanisms involved in the operation of the social macrocosm are energy-consuming devices, therefore the basic metrical relationships are: the factor of energy conversion, or efficiency; and the rate of conversion of available energy of the mechanism as a functional whole in a given area per time unit. Technocracy accordingly establishes a new technique of social mensuration, that is to say, a process for determining the rates of growth of all energy-consuming devices within the limits of the next most probable energy state.

The Energy Survey of North America now being conducted by Technocracy in association with the Industrial Engineering Department of Columbia University and the Architects' Emergency Committee has found that employment of this method has not only yielded new data but has endowed already existing data with a new significance. As the above method is one of measurement, it follows axiomatically that all processes of evaluation are excluded. Value has no metrical equivalent.

Value is defined by the economists as the measure of the force of desire. It has its physical manifestation in any one commodity unit by which all other commodities or services are evaluated. Any society using a commodity method of valuation shall herein be said to be employing a price system. A "social steady state" is a social mechanism whose per-capita rate of energy conversion is not changing appreciably with time. Social change, on the other hand, may be defined as the change in the per-capita rate of energy conversion, or the change from one order of magnitude to another in the social conversion of the available energy. All social history prior to the last century and a half, viewed technologically, may therefore be described as the record of a steady state. Only within the last hundred and fifty years has there been introduced a technique that has specifically caused social change. Technology, as the executor of physical science, is the instrument for effecting social change.

During the 200,000 years prior to 1800 the biological progression of man, in his struggle for subsistence on this earth, had advanced so far that the total world population in that year reached the approximate number of 850,000,000. During the subsequent 132 years world population has attained such heights that it now exceeds a total of 1,800,000,000; in other words, the population increase in the last 132 years has been greater than it was in the previous 200,000. Most of this increase in the human species has been made possible by the social introduction of technological procedures, that is, change in the means whereby we live as brought about solely by the introduction of technology.

II

A century ago these United States had a population of approximately 12,000,000, whereas to-day our census figures give a total of over 122,000,000—a tenfold increase in the century. One hundred years ago in these United States we consumed less than 75 trillion British thermal units of extraneous energy per annum, whereas in 1929 we consumed approximately 27,000 trillion British thermal units—an increase of 353 fold in the century. Our energy consumption now exceeds 150,000 kilogramme calories per capita per day; whereas in the year 1800 our consumption of extraneous energy was probably not less than 1600 or more than 2000 kilogramme calories per capita per day.

The United States of our forefathers, with 12,000,000 inhabitants, performed its necessary work in almost entire dependence upon the human engine, which, as its chief means of energy conversion, was aided and abetted only by domestic animals and a few water wheels. The United States to-day has over one billion installed horse power. In 1929, these engines of energy conversion, though operated only to partial capacity, nevertheless had an output that represented approximately 50 per cent of the total work of the world. When one realizes that the technologist has succeeded to such an extent that he is to-day capable of building and operating engines of energy conversion that have nine million times the output capacity of the average single human being working an eight-hour day, one begins to understand the significance of this acceleration, beginning with man as the chief engine of energy conversion and culminating with these huge extensions of his original one-tenth of a horse power. Then add the fact that of this 9,000,000 fold acceleration 8,766,000 has occurred since the year 1900.

Stated in another way, if the total one billion installed horse power of the United States were operated to full capacity, its output would be equivalent to the human labor of over five times the present total world population.

Physical science has outdistanced present social institutions to such an extent that man, for the first time in history, finds himself occupying a position in which a complete utilization of his knowledge would assure the arrival of certainty in a continental social mechanism. Man, in his age-long struggle for leisure and the elimination of toil, is now at last confronted not only by the possibility but by the probability of this arrival. Such a new era in human life is technologically dependent only upon an extension of the physical sciences and the equipment at hand.

But the pathway to that new era is blockaded with all the riffraff of social institutions carried over from yesterday's seven thousand static years. The law of the next arrival is depicted by the Gaussian curve of probability, or the next most probable energy state.

III

America faces the threshold of the new era with the greatest total debt load ever carried by any social mechanism, a debt of over $218,000,000,000 against her physical equipment and its operation. With the number of unemployed greater than the total population of a century ago; with one of the most providential geologic set-ups of any continental area; still possessing more energy and mineral resources than any like area on the world's surface; having more than one billion installed horse power of prime movers wherewith to degrade available energy into use-forms; possessing a personnel of over 300,000 technically trained men in many varied engineering fields and more than 4,000,000 men partially trained and functionally capable of operating the greatest array of productive equipment ever at the disposal of man—with all this, we have, nevertheless, failed to profit from technological advances, and accordingly find ourselves, for the first time in history, with an economy of plenty existing in the midst of a hodge-podge of debt and unemployment.

America can expect no help in the solution of this problem from any current social theory. What has the world to offer toward such a solution? Europe discovered America in 1492. To-day America is further away from Europe than she was when Columbus sailed. The America of to-morrow will necessitate a rediscovery by Europe. European culture and traditions have nothing of worth-while importance to offer America in this twilight period preceding the dawn of a new era. No European importations of social or political theory can have the slightest value in solving the operational problems facing America to-day. Arising out of areas that lack adequate physical equipment and trained personnel, areas in which only a low percentage of the population is disciplined in engineering thought processes, European socio-political philosophies and theories are the natural outgrowth of a more classified division and orientation of the entrepreneur sectionalism of the price system. No theory of social action or governance now existing or proposed in Europe would in any way be endemic to that unique set-up of geologic conformation, technique, equipment, and personnel peculiar to North America.

Russia, of whose population 92 per cent were tillers of the soil under the *ancien régime* and which had meagre technical facilities and more musicians than technologists, found itself in the position of being compelled to inaugurate an industrial era under a communistic price system of production. Soviet Russia was forced to call upon the outside world for technical assistance in order to perpetrate reproductions of factories already obsolescent from an obsolescent price system. Russia, in its Parthian retreat from capitalism, has scored but a Pyrrhic victory. It mistook the name tag of one phase of the price system for that system's entirety; it abandoned the tag, but retained the essential mechanics.

To approach social phenomena by substituting Hegelian for Aristotelian dialectics may be an interesting intellectual pastime, but it has no functional importance: it is but one more recrudescence of the philosophic futility implicit in European tradition.

IV

The England of the Black Prince, with its population of 5,000,000, its wealth of oak timber, its hearty people drinking deeply of ale (made not from hops but from barley malt), its original resources of copper, lead, tin, iron ore, and coal—this England developed under the price system of production. Inevitably, like the prodigal son, England went forth into the world and squandered its inheritance among the harpies of world trade and debt creation.

The United Kingdom, with an area of 94,000 square miles and a population of 46,000,000,—or a density of 490 inhabitants per square mile,—with arable land amounting to only 23 per cent of the total national area, finds itself in the physical position of possessing only a single energy resource, and that a declining one. Its tin gone, as well as its copper and lead, its iron requiring 56 per cent foreign benefication in order to produce steel, its coal becoming more and more difficult to mine, the United Kingdom is fast retrogressing from its position as the possessor of easily available energy to its next most probable energy state as two islands off the coast of the European continent. A valiant race, fighting a losing battle, is displaying an admirable fortitude in the crisis that is resulting from excess population, declining resources, and obsolescent equipment operated by the antiquated methods of a price system.

The United Kingdom will be forced by internal pressure to adopt measures even more extreme than the flight from the gold pound. It may be compelled by the growing disparity between its own industrial operation and the world balance to such extremities as abandonment of monetary currency and the accompanying credit structure. In that event, a British currency of pure fiat power might be attempted as a last desperate resort. The present deflationary programme may be reversed in the near future to one of inflation, a last straw grasped at in England's struggle for the export markets of the world. Sooner or later, in spite of British imperialism, the

United Kingdom, under a price system, will be forced to meet a situation that will be increasingly grave in its internal operation. There remains only the colonizing soporific of bestowing a surplus population of 35,000,000 on the oversea Dominions.

Fascism, that strange but natural partnership of the Italian political state and vested interests, is a process of consolidating all the minor rackets into one major monopoly. Such a condition brought with it the sequelæ of discipline and sanitation that necessarily accompany complete trustification. Italy, which is insufficiently supplied with energy and mineral resources, which possesses only a limited amount of water power and volcanic heat, which has some mercury and sulphur but no coal, oil, or gas, no iron ore, copper, tin, lead, or zinc, and which lacks a high enough percentage of arable land to grow sufficient foodstuffs for its own needs—Italy belongs to the geologic order of areas that can not create and operate an industrial energy civilization. Fascist Italy is rapidly increasing its dangerous overload of population by granting national bonuses to large families in furtherance of its *mare nostrum* policy. Fascism is an attempt at a last-ditch defense of a price system, an effort to maintain an unbroken front against oncoming social change, but this unbroken front is spurious in that it is being temporarily maintained by foreign importation of energy-resource materials, supplemented by the manna of the Lord.

Egypt, Assyria, Greece, Rome, and, in the Victorian age, Imperial Britain have all led the world in their day; each in turn has been the vanguard of civilization. The past is strewn with ruins of empire. Now there is but one continental area that from the standpoint of its geologic set-up, equipment, personnel, and the state of its technology is competent and ready to inaugurate a new era in the life of man.

V

America stands on the threshold of that new era, but she will have to leave behind all the wish-fulfilling thought and romantic concepts of value that are the concomitants of a price system. So, too, all philosophic approaches to social phenomena, from Plato to—and including—Marx, must functionally be avoided as intellectual expressions of dementia præcox. Economics, that pathology of debt, not containing within itself any modulus or calculus of design or operation, must likewise be discarded with the other historical antiquities. No political method of arriving at social decisions is adequate in continental areas under technological control, for the scientific technique of decision arrivation has no political antecedents.

Under a price system wealth arises solely through the creation of debt. In other words, price-system wealth consists of debt claims against the operation of the physical equipment and its resultants. Physical wealth, on the other hand, is produced by converting available energy into use-forms and services. The process of being wealthy is the degradation of the resultants of the above conversions into complete uselessness—in other words, total consumption. To be physically wealthy is not to own a car but to wear it out. Technology has introduced a new methodology in the creation of physical wealth. It is now able to substitute energy for man hours on the parity basis that 1,500,000 foot pounds equals one man's time for eight hours. National income under the price system consists of the debt claims accruing annually from the certificates of debt already extant. Physical income within a continental area under technological control would be the net available energy in ergs, converted into use-forms and services over and above the operation and maintenance of physical equipment and structures of the area.

Individual income under a price system consists of units that are not commensurate with the quanta by which the rate of flow of the physical equipment is measured and upon which the social mechanism depends for its continuance. Individualism is therefore favored under a price system, since individualism can obtain a monetary equivalent proportional to the individual's ability to create debt. Individual income under such a system therefore depends on the extent to which advantage is exercised by means of the interference-control that is dominant throughout the whole system of debt creation.

Individual income under technological control would consist of units commensurate with the quanta by which the rate of flow of the physical equipment is measured throughout the entire continental area. The unit income of the individual would be determined by the period necessary in that area to maintain a thermo-dynamically balanced load, that is to say, the time it takes for a complete cycle of the operating and production procedures to be completed.

Any unit of *value* under a price system is a certification of debt. Any unit of *measurement* under technological control would be a certification of available energy converted. Such units of certification would have validity only during the balanced load period for which they were issued. This method of producing physical wealth and measuring its operation precludes the possibility of creating any kind of debt. It also eliminates the entire domain of philanthropy. Furthermore, all bonds, financial debentures, and other instrumentalities of debt would cease to exist, since they do not have one iota of usefulness in the physical operation of such an area under technological control.

Technocracy proposes no solution, it merely poses the problem raised by the technological introduction of energy factors in a modern industrial social mechanism. Continental America possesses all the essential qualifications for such a mechanism —sufficient energy and mineral resources, adequate water precipitation, more than enough arable land of proper chemical stability, highly developed technological facilities backed by a trained personnel, powerful research organizations. All these things are entirely sufficient to assure the continuance of a high energy standard of livelihood for at least a thousand years, if they are operated on a nonprice basis with the technological means known at present.

America stands now at the crossroads, confronting the dilemma of alternatives. The progression of a modern industrial social mechanism is unidirectional and irreversible. Physically this continental area has no choice but to proceed wtih the further elimination of toil through the substitution of energy for man hours. There can be no question of returning to premachine or pretechnological ways of life; a progression once started must continue. Retrogressive evolution does not exist.

INTRODUCTORY NOTE

The Technocrats liked to call Thorstein Veblen (1857–1929) a founder of their movement. In 1919–1920, Veblen, an unorthodox American economist and critic of social and economic institutions, wrote a series of articles that were later collected and published as *The Engineers and the Price System.* In these he explored the role of the businessman, the manager, and the engineer in the American economy and society. Veblen deplored the way in which profit-motivated businessmen and financiers tolerated, even cultivated, inefficient production in order to increase profit. He believed that these diseconomies would be eliminated and production increased if engineers took over the management of the American system of production and applied to it their objectivity, rationality, and methodology. For a brief period, around 1920, Veblen and a group of reform-minded engineers attempted to forge an alliance, but with slight success, for Veblen wanted a radical reorganization of the social and economic structure and the engineers then wanted a larger role in the existing one.

Because of Veblen's emphasis on rational procedures and maximum production and his dislike of the price system, the Technocrats of Howard Scott's movement found Veblen's views congenial. But, as Leon Ardzrooni, an engineer who had been close to Veblen points out in "Veblen and Technocracy," the Technocrats of the early thirties did not explicitly call for the management of the society and the economy by engineers that Veblen had seen as a *sine qua non* for society fully to enjoy the benefits inherent in modern production technology.

Leon Ardzrooni was a friend and colleague of Thorstein Veblen at the New School for Social Research in New York.

Veblen and Technocracy

LEON ARDZROONI

I

The name "Technocracy" has been in everyone's mouth of late and men everywhere have turned to it with breathless expectation for the solution of the manifest economic paradoxes and inconsistencies that now beset the nation. The anxious acclaim with which this proposal has been received in many quarters merely attests to the all-pervading fact of the well-nigh complete derangement and confusion of the industrial forces of the country. To further enhance its prestige, Technocracy has been surrounded with the halo of the great name of Thorstein Veblen. What has lent color to such an authentication has been the second printing recently of Veblen's *The Engineers and the Price System,* which the critics have hailed as "the original gospel from which the theories of Technocracy have been developed."

In what manner and to what extent is this true? What was Veblen's relation to the scheme that is now called Technocracy? In the light of his social and economic theories, what would Veblen have thought of it? These are some of the questions which, in view of the current excitement and agitation concerning Technocracy, become of considerable human interest.

During the year 1919, when Veblen found himself out in the cold world, foot-loose and jobless, he was retained at a modest annuity by *The Dial,* then in the control of Miss Helen Marot and Mr. Martin Johnson. In his capacity as a regular contributor Veblen wrote a number of essays and articles for that magazine which were subsequently (1921) collected and published by Mr. Huebsch under the title, *The Engineers and the Price System.*

In the meantime Veblen had been given a comfortable berth at the New School for Social Research, which for a

The Living Age, 344 (March, 1933), 39–42.

time achieved undue notoriety as a radical institution, but which has since succeeded in living down that ignominy. Here Veblen conceived the idea of getting together a group of like-minded folk chosen from among young economists, accountants, engineers, and technicians generally to form the nucleus of a "soviet of technicians," a brief working plan for which was contained in *The Engineers and the Price System*. In due time the New School came to be, informally, the headquarters for such a group, which, with the exception of two or three members, was composed of persons not connected with the New School. Among these the most outstanding was Howard Scott, who recently has come forward as the titular head and official spokesman of Technocracy. Veblen took an unusual interest in Scott because, as a highly trained technician,—a type that Veblen respected and admired,—he seemed to fit in well with Veblen's scheme of things and also because Scott's experiences as an engineer during the War, as he related them, were unusually edifying and instructive.

Veblen's health did not permit him to indulge in long-drawn-out and frequent conferences and discussions with the group, but a few of the younger men spent considerable time with Scott. Later on, at the earnest solicitation of Veblen and at some expense to the New School, a prominent and experienced engineer joined the group, chiefly for the purpose of consulting with Scott. The result of these consultations and conferences, it is regrettable to say, was not very satisfactory, largely because Scott remained somewhat of an enigma to most of the group. But Veblen maintained his faith and interest in Scott to the end.

At about this time (1921) the New School became involved in financial and other difficulties, with the result that the group became more or less disorganized, and Veblen, because of his failing health, withdrew from activities of that nature, though he continued to lecture at the New School.

II

From this brief narrative it should be clear that Veblen had laid the foundations and worked out the details of what passes current as Technocracy before he came to the New School and before his contact with Howard Scott. As for the basic idea in

Technocracy, in so far as that centres about the repudiation and rejection of bankers, financial agents, statesmen, and so on, as nothing better than marplots in the conduct of the productive and distributive processes of industry, that idea runs through practically all the important works of Veblen.

The immediate occasion, however, for Veblen's active interest in what is now called Technocracy was his experiences in the Food Administration and his observations of the queer and incredibly imbecile antics of the dollar-a-year men in Washington during the Great War, not to mention the untold stupidities and skulduggeries in other sections committed in the name of God and country. After the premature and abortive end of the War, in his frequent discussions of current problems, Veblen often expressed the firm conviction that things could not continue much longer under the guidance and management of the vested interests and that sooner rather than later a change was unavoidable. The problem that therefore engrossed his attention was as to how the inevitable change might be brought about without a serious dislocation and disintegration of industry and consequent hardships and sufferings for the underlying population. To the solution of this problem he addressed himself in the series of articles that now compose *The Engineers and the Price System.*

The plan as outlined in those articles differs radically from that envisaged by the present-day spokesman for Technocracy. In the first place, Technocracy excludes all but technicians from the control of industry. In the second place, it is silent as to the manner and method of securing that control.

With respect to the first of these points, according to Veblen the technicians, or, as he sometimes called them, the "production engineers," can not be counted on to set up the new régime alone and by themselves but must be aided by "consulting economists." Here it should be noted that Veblen is careful to point out that he does not refer to the "certified economists" who "by reason of doctrinal consistency and loyalty to tradition have habitually described business enterprise as a rational arrangement for administering the country's industrial system and assuring a full and equitable distribution of consumable goods to the consumers. . . . Quite blamelessly, the economists have, by tradition and by force of commercial pressure, habitually gone in for a theoretical inquiry into the ways and means of salesmanship, financial traffic, and the distribution of income and property, rather than a study of the industrial system considered as a ways and means of producing goods and services. Yet there now are, after all, especially among the younger generation, an appreciable number, perhaps an adequate number, of economists who have learned that 'business' is not 'industry' and that investment is not production. . . . 'Consulting economists' of this order are a necessary adjunct to the personnel of the central directorate, because the technical training that goes to make a resource engineer or a production engineer is not of a kind to give him the requisite sure and facile insight into the play of economic forces at large."

In this connection students familiar with Veblen's earlier writings will readily call to mind his critical essays on economic theory. These strictures, however, on the guild of economists are particularly timely in view of the appointment in recent weeks of several committees in different parts of the country composed of the "best minds" to deal with the present economic debacle. With regard to the competency of the "best minds" to cope with the elements in a social crisis it may appear as a singular coincidence that Lenin also, be it recorded to his everlasting fame, abjured all relationship with the "best minds." This was the issue upon which the Mensheviki and the Bolsheviki parted company. That one stroke of genius on the part of Lenin perhaps more than any other single factor decided the course of events and the fate of the revolution in Russia.

With regard to the manner and means of securing control of industry, there are two main lines to be followed by the soviet of technicians:

> (a) An extensive campaign of inquiry and publicity, such as will bring the underlying population to a reasonable understanding of what it is all about; and (b) the working out of a common understanding and a solidarity of sentiment between the technicians and the working force engaged in transportation and in the greater underlying industries of the system: to which is to be added, as being nearly indispensable from the outset, an active adherence to this plan on the part of the trained workmen in the great generality of the mechanical industries.

Veblen, with his uncanny insight, foresaw the "danger of a revolutionary overturn" and clearly set forth "the circumstances that make for a change." He also offered a workable plan to effect the change without inflicting undue pain and privation on the underlying population. The plan is workable, of course, just so soon as the underlying population gets to thinking along these lines and precipitates "the abdication of the Vested Interests under conviction of total imbecility." But "the underlying population can be counted on stolidly to put up with what they are so well used to, just yet," and "the Vested Interests are secure in their continued usufruct of the country's industry, just yet" and "so long as there is no competent organization ready to take their place and administer the country's industries on a more reasonable plan."

III

As matters stand to-day Technocracy offers no practicable remedy for the chronic ills that afflict the economic life of the nation. It has neither the definitive and convincing character nor the bold and revolutionary implications of Veblen's "soviet of technicians." In fact, there is nothing in the proposed scheme of Technocracy to prevent it from being used as an instrument of oppression and exploitation of the underlying population quite as merciless and relentless in its incidence as any that exists to-day. The most sinister and menacing feature of the proposal is that contained in the implication of the domination of unskilled and unblessed workmen by their skilled and more fortunate brethren. And now that Technocracy has enlisted the active interest of the "best minds" in its behalf it may confidently be expected that presently the "Guardians of the Vested Interests" will bestir themselves to seize the proffered opportunity to bring about a change in the present unhappy conditions without altering them.

So that, in the words of Veblen written fourteen years ago, "there is nothing in the situation that should reasonably flutter the sensibilities of the Guardians or of that massive body of well-to-do citizens who make up the rank and file of absentee owners, just yet."

INTRODUCTORY NOTE

"Technocracy—Boon, Blight, or Bunk?" summarized opinion for and against Technocracy. The supportive material (selected from various journals) attributed unemployment, the nation's greatest problem, to the failure of society's political and economic arrangements to exploit fully the potential of modern technology—a situation Technocracy was designed to end. The arguments against Technocracy emphasized the incomprehensible jargon of Scott and his group and questioned his basic statistics on the increase of productive capacity. Even critics, however, acknowledged the obvious inability of society to utilize fully the technology of production and distribution. But, having acknowledged this, they then recommended or anticipated solutions less radical than Technocracy. They foresaw, for example, the rising expectations of Americans for goods and services and the needs of the underdeveloped nations absorbing the productive capacity of the country.

Technocracy—Boon, Blight, or Bunk?

Technocracy is all the rage.

All over the country it is being talked about, explained, wondered at, praised, damned.

It is found about as easy to explain—at least by the average editor—as the Einstein theory of relativity.

Perhaps that is because it is a combination of several things. The basic idea now being advanced by its spokesmen is that the advance of the machine age is forcing overproduction and underemployment at such a rapid, accelerating rate that our industrial civilization with its tremendous burden of money debt is bound to collapse within a very few years unless we change the whole scheme of things. And this doesn't mean merely Socialism or even Communism. In the words of Howard Scott, whom we may call the Lord High Technocrat, it means complete abolition of the "price system," which would include the end of all banks, all philanthropy, "all bonds, and other instrumentalities of debt."

In a narrower sense, Technocracy, to quote Mr. Scott again, "is a research organization, founded in 1920, composed of scientists, technologists, physicists, and biochemists." This group, of which Mr. Scott is the leader, "was organized to collect and collate data on the physical functioning of the social mechanism on the North American continent, and to portray the relationship of this continent and the magnitude of its operations in quantitative comparison with other continental areas of the world."

Technocracy, explains Mr. Scott in his *Living Age* article, which he calls the only official statement, is establishing a new technique in social mensuration, "that is to say, a process for determining the rates of growth of all energy-consuming devices within the limits of the next most probable energy state." Mr. Scott is generally credited with originating the new "theory

The Literary Digest, (December 31, 1932), 5–6.

of energy determinants," altho press investigators have traced the basic idea back to the late Thorstein Veblen.

The Technocracy group are now carrying on what they call an "Energy Survey of North America," which has been given room in Columbia University, and is furnishing employment to a number of otherwise jobless draftsmen.

The Technocracy argument is that technology has so outdistanced our whole political, economic, business, and social set-up that the only salvation is to scrap everything and turn the control over to the technologists. There will be an entirely new unit of value based on energy consumption. "Any unit of measurement under technological control would be a certification of available energy converted." But when it comes to explaining how the change is to be made, Mr. Scott becomes rather more vague than usual. As he puts it: "Technocracy proposes no solution, it merely poses the problem raised by the technological introduction of energy factors in a modern industrial social mechanism."

But it is explained by this gentleman in a *Harper's Magazine* article that we are standing on the threshold of "what is simultaneously opportunity and disaster." On the disaster side, "the mills of the gods have ground almost their allotted time, and they have ground exceedingly fine"—

> The spectacle of a New Jersey rayon factory now being designed to run eventually without human labor, save for one man at the switchboard, is more than a warning of further unemployment, more than a notice to competitors that a rival has lowered his production cost to a minimum.

It is proof, in Mr. Scott's modest way of putting it, that "the bankers, the industrialists, the Marxists, the Fascists, the economists, the soldiers, and the politicians are things of the past."

This looks pretty serious, especially when bolstered up with scores of instances and reams of figures showing the tremendous rate at which machines are displacing workers. Some of the interpreters of Technocracy predict 20,000,000 unemployed next winter, and the complete collapse of civilization in the spring of 1934.

Here is the cheerful side. As the Technocracy experts are quoted by Wayne W. Parrish in *The New Outlook:* "Man in his age-long struggle for leisure and the elimination of toil has finally arrived at that position where for the first time this goal is not only possible but probable." It is pointed out that in such a complete self-sufficient area as North America, "with what is known of technology to-day in this country, it is now necessary for the adult population, ages twenty-five to forty-five, to work but 660 hours per year, per individual, to produce a standard of living for the entire population ten times the average income of 1929." But we can not attain this golden age of leisure without scrapping our present economic and financial machinery. "Technology has written 'mene, mene, tekel, upharsin' across the face of the price system."

Without ever once mentioning the word Technocracy, the magazine *Fortune* discusses this problem of "obsolete men." It sees the new power revolution about "to replace man permanently as a source of energy, and to install him in a new and limitable function as a tender of machines." Whereas it has been assumed that men thrown out of work by labor-saving machines can find employment sooner or later in new industries, we are now, in the words of the Secretary of Labor in 1927, "developing new machinery at a faster rate than we have been developing new industries." And while the spread-work plan is helping in the present emergency, "as a permanent device it is evident that if the rate of mechanization continues to rise, as it is now rising, a time will shortly be reached when the rationing of the residuum of work among the total number of wage-earners with corresponding cuts in real wages would result in no wage-earner receiving a living wage."

Perhaps that is enough to give an idea of the assertions that are being echoed or challenged in the editorial columns of practically every newspaper in the country. *The Nation* calls it all "the first step toward a genuine revolutionary philosophy for America." The president of the Case School of Applied Science in Cleveland says that certain of Mr. Scott's facts may be wrong, "but the total picture is substantially true."

And now for the critics of Technocracy.

Let us hear first from such an authority as Dr. Walter R. Ingalls, president of the American Institute of Weights and Measures:

> Convalescing business has been hit below the belt by a pseudo-scientific fist, armed with the brass knuckles of an imposing but meaningless scientific jargon. Technocracy has presented a multitude of alarming conclusions, but has withheld supporting facts.

The sharpest challenge of all, perhaps, comes from John H. Van Deventer, editor of *The Iron Age*. First he hits the Technocracy statistics. As an authority on the iron and steel business, he says flatly: "Instead of productivity in pig-iron having increased 650 times in fifty years, as technocracy claims, it has increased 23.2 times." Mr. Van Deventer declares that the technocracy spokesmen have similarly exaggerated the acceleration of technological unemployment in other lines. Then he gives figures of his own to show that instead of the machine driving the people out of work, it makes work for them:

> In 1930, we find the density of employment greater instead of less after thirty years of our most intensive mechanization. For there were then 398 breadwinners for each thousand of our population as contrasted with 383 in 1900. A net gain of fifteen workers per 1,000 of population during the thirty-year period.

Such figures should prove conclusively, in Mr. Van Deventer's opinion, "that machinery, with its continuous record of multiplying working opportunities, is not the cause of the depression." He quotes approvingly this statement by the New York *Times:*

> The glut in primary commodities is not the result of too many machines, but of too many acres of coffee, rubber, beef, wool. It is not the result of technology, but of human miscalculation. For nearly two hundred years the machine has been operating to enrich the existence of the world's masses. It is silly, on the basis of three years of hard times, to announce that the machine has suddenly become man's destroyer.

Some critics are ironic. "All that the elaborate technical rumble-bumble of the 'Technocratic' statistics shows is the fact that by the use of power machinery men can produce much more per man hour than they could by hand labor, especially in the United States," according to *The Business Week,* which wonders what there is new about that. But the business man, we are told, "disturbed and puzzled by the difficulties of a prolonged depression, is unprepared when the Technocrats mumble some obscure mumbo-jumbo about 'energy determinants' and 'decision arrivation' over him, pronounce the doom of the 'price system' with the portentous finality of a papal bull itself, and sentence him and business to death."

The Technocracy people, continues *The Business Week*, ignore the actual increase in our working population, the shifts in occupation created by new demands of new industries—

> They say nothing about the constant demand for labor created by obsolescence of business methods, industrial processes, products, and equipment, under pressure of changing social standards of living, or the constant shift of workers from the mechanical industries to the service businesses—nothing except to condemn it all as wasteful of natural resources.
>
> They say nothing about the possibilities of raising standards of living and demand for industrial products in undeveloped countries, except to dismiss it as impossible because there is not energy enough available—a strange thing for engineers and scientists to say when they know that any day may see new sources of energy opened to exploitation.
>
> Finally, they say that whatever substitute is devised for the price-and-profit system, it must be one based upon scientific control, planned on a national scale and operated by technical experts. Human nature must have nothing to do with it; human desires, human behavior must not be allowed to interfere. It is too complex, too delicate, too speedy to be bothered by mere human beings. The laws of physics must be the sole and supreme law of the land. The system of government must be designed by scientists and engineers, and administered by a kind of Technocratic National Committee.
>
> What they will do about politics, fashion, fun, and other persistent human dispositions and frailties remains to be seen. Probably just ignore them."

All that Mark Sullivan of the New York *Herald Tribune* can see in technocracy is "one of those novel ideas with a mystic sounding name which sweep over the country once in so often."

To the editor of *Barron's Weekly*, Mr. Scott's argument is only "a shade more intelligible than would be a similar amount of space fully occupied by a series of Einsteinian equations."

But a number of the critics of Technocracy admit that it has called attention to an extremely serious situation, and one which requires all our energies to find a cure. Let us take, for instance, the carefully worded statement given to the United Press by such a business authority as James D. Mooney, President of General Motors Export Corporation:

> Technocrats who hold that a breakdown would be inevitable if the 1929-1932 ratio of decline were maintained for another two years may be right.
>
> Much remains to be done. It is obvious that we are to-day suffering from a one-sided deflation. Salaries, wages and commodity prices have been deflated, but many groups of values remain at inflated levels.
>
> The price and rate structure are to-day still out of balance. The internal structure urgently requires deflation, a situation that can be met either through inflation, revalorization or by government decrees, the third course seeming the soundest.
>
> In the field of international politics, closely interwoven with economics, the entire debt and trade situation requires reestimation and clarification.
>
> If I believed the American ingenuity, honesty and individual courage were not able to cope with these last remaining stages of deflation I should feel, with the gloomy philosophers of Technocracy, that a breakdown was inevitable. But every indication points to the solution of these problems, and every business leader and economist of the country is intent upon their solution.
>
> It is entirely reasonable that at such a time, doubting and fear-struck people would react to prophecies of doom due to mechanization. Such prophecies have been consistent since 1830. Inevitable dark days have passed, to be succeeded by days in which man's conquest and utilization of the machine led to a busier, brighter, happier world, with more employment for all, more possessions, more basic values.
>
> The price system, which "technocracy" holds untenable, is the bedrock of the American economic scheme. It is the key which has unlocked the wealth inherent in machines, has made us masters of machines for more than one hundred years, has made machines give us increasing values and utilities.
>
> The breakdown of 1929 was not a mechanical breakdown. It was caused by a lack of understanding and respect for economic laws.

And the New York *Journal of Commerce* offers this advice:

> Better credit control and encouragement of more even popular spending habits through tax policies and voluntary avoidance

> on the part of business management of excessive resort to instalment selling, unbalanced plant expansion, etc., would help. Adoption of more logical foreign policies, involving either consistent isolation or consistent freedom of trade and financial intercourse with foreign countries would also facilitate stabilization.
>
> Specific remedies of this kind should be evolved and tried before serious consideration of anything more revolutionary.

The Technocrats may be wrong in their figures and may be overstating their case, but the Baltimore *Sun* admits that "the adaptation of our economic machinery to our increasingly efficient industrial machinery on generally beneficial economic terms is so obviously a pressing national concern that the Technocrats have performed no slight service in devising a means of directing animated public attention to it."

We are a long way from the abandonment of a price system and a money economy; however, the Cleveland *Press* concludes: "We are not far from the date when shorter hours throughout industry, the diversion of more of the fruits of the industry to the workers, unemployment insurance, old-age pensions, some kind of economic planning, some kind of social control of the business cycle, are imperative." And to let the St. Louis *Star* have the last word, "it is just as safe for the United States to ignore technological unemployment as it is to fire a boiler without adding water to it."

INTRODUCTORY NOTE

Archibald MacLeish sensed that Technocracy, though discredited at the time he wrote, spoke to a fundamental issue. Technology, he realized, had changed the material base of Western civilization: in principle, all men could "work less and enjoy more." Technological unemployment and poverty amid immense productive capacity, he believed, were the negative outcome of the failure of society to fulfill the positive potential of technology. In "Machines and the Future," MacLeish acknowledged that Scott and his movement had been discredited by faulty tactics and facts exploited by their foes, but the discrediting was, in his view, superficial. The critical problem to which Scott addressed himself remained and demanded, in MacLeish's opinion, a response of the order and magnitude of Scott's Technocracy. "Those who ignore the problem and those who discredit the hope," MacLeish concluded, "do so at their peril."

Archibald MacLeish (1892–   ), an American poet and essayist who served as Librarian of Congress from 1939 to 1944, has been awarded three Pulitzer Prizes for his poetry and drama. Among his essays are articles on housing in America and works on American values and history. In 1929 he was in Mexico with the staff of *Fortune* magazine, and his Pulitzer Prize-winning narrative poem, *Conquisitador,* was drawn from experiences there. He has also taught poetry and creative writing at Harvard University.

Machines and the Future

ARCHIBALD MacLEISH

Nothing about the human race is more curious than its passion for believing prophecies of doom—unless it be its passion for destroying the prophets. Any oracle who cares to foretell the end of a civilization can make the front pages from Bushire to Nome. But God must defend him if the details of his divination err.

Mr. Howard Scott, augur of the Technocrats, is the latest example. Like all successful prophets Mr. Scott appeared in a time of tribulation. Like all successful prophets Mr. Scott foretold woe. And like all successful prophets Mr. Scott was widely and resoundingly and hysterically believed. With the inevitable reaction, the inevitable repudiation, and the inevitable relapse. Indeed, Mr. Scott differs from his predecessors among the ancients in one characteristic only. The Roman prophet could not be discountenanced until the event. But Mr. Scott, because he appealed to the great American belief in scientific truth and therefore expressed his dooms in scientific terms supported by scientific findings, was subject to immediate check. And when the check was made, when the devout discovered that the ratio of 1932 electric-bulb production per man to 1914 electric-bulb production per man was 550 to 1 instead of 9,000 to 1, the black ink of rage ran from one end of the country to the other. The customers had been sold. The *facts* were otherwise. Mr. Scott was a fake prophet. And nothing was necessary to a proper cinematic ending for the whole affair but the repudiation of Mr. Scott by his own disciples. Which obligingly followed.

If no one were concerned in the result but Mr. Scott, the situation would be innocuous and amusing. Mr. Scott would have had his day in the sun and no one but himself to blame for the precipitancy which threw him into the newspapers

The Nation, 36 (February 8, 1933), 140–142. Reprinted by permission.

before he had tested his statistics. Unfortunately, however, Mr. Scott is very far from being the only person concerned. For his name, in part because of his own egoism, and in part because of the fumbling of editors who knew no word to express the corpus of ideas with which Mr. Scott was dealing, has become identified and synonymous with one of the most pressing problems our civilization has to face. Technocracy, whether we like it or not, now stands by substitution for the whole body of social and economic questions posed by the mechanization of industry. And the neglect of Mr. Scott as an individual and of Technocracy as a school means the neglect of issues which we cannot afford to neglect. As the published remarks of the last two weeks abundantly prove. It is not alone left-wing intellectuals, recently converted from Professor Babbitt to Karl Marx and unable with decency to afford a second removal at this time, who have breathed audible sighs of relief over Mr. Scott's discomfiture. It is also the bankers and the episcopal bishops and the societies of engineers and the chambers of commerce and the Professor Fishers. Mr. Paul du Pont put the case neatly when, with unintended irony, he remarked that United States Steel at sixty would put an end to the whole discussion. It would. And these gentlemen hope to see the day.

The rest of us, however, do not. And for two reasons. Of which the lesser is the certainty that failure to deal with the problem now means a disaster—and a disaster beyond the profit of any revolutionary group—within this generation; and of which the greater is the conviction that the recognition of the technological trend and the acceptance of its significance offer the first ray of human hope since industrialism began. The Technocrats were journalistically clever but humanly obtuse when they made their cause a cause of fear. The real meaning of modern industrialism has no interpretation in terms of fear except for the banking class. For the rest of society it offers hope. Earlier in this century and throughout the latter half of the nineteenth century there was no hope in industrialism but escape. Poets turned aside from it. Philosophers transcended it. Even the Communists who plunged into it to the hair roots, accepting the human categories of economic class it imposed, worshiping the symbolic tractors it produced, and making a religion of the labor it compelled men to perform, were unable to find in industrialism a positive justification and were forced to express their Utopia in negatives—the *end* of capitalism, the *end* of exploitation, the *end* of injustice. But with the third decade of the present century the image changed. Industrialism turned another face. It became apparent that the goal of industrialism was not merely greater and greater production, more and more material, richer and richer bankers, but something quite different from these things. It became apparent that industrialism was moving toward a degree of mechanization in which fewer and fewer men need be, or indeed could be, employed. And that the result of that development must, of physical necessity, be a civilization in which all men would work less and enjoy more. For the alternative to such a choice was a state in which a half or a quarter of the adult population would be unemployed and carried as charity charges by the remainder of the population, and such an alternative was unthinkable. What political or social revolutions must intervene no man could say. But the probable and indicated outcome could be foreseen by any man who was familiar with the figures.

The statistical statement of that prognosis is the heart of Technocracy. And it is there, rather than on the pin's head of particular production figures, that the angels should be made to dance. Technocracy's brick and pig-iron figures may, and undoubtedly do, prove that the Technocrats are exceedingly careless engineers. But in the last analysis the important question is not whether a man can make 9,000 times or only 550 times as many light bulbs now as he could make in 1914, but whether it is or is not true that the mechanization of industry has reached the point where production can increase without a corresponding increase in the number of wage-earners. For if the answer to that question is in the affirmative, then the direction of growth of industrialism has changed, our civilization has turned a corner, and the ancient conception of human work as the basis of economic exchange and of the right to live is obsolete, since the work of machines and the conversion of non-human energy take the human place.

There is no doubt whatever that the answer is in the affirmative. Charts in the preparation of which the Technocrats had no part prove that from and after the year 1919, while the volume of production increased, the number of wage-earners employed in manufacturing leveled off and actually fell, with a loss by 1927 of 3 percent. And this loss was reflected in actual unemployment totals. Dr. Leo Wolman, whose findings are not subject to the skepticism which meets Mr. Scott's, estimated a minimum of 2,055,000 unemployed outside of agriculture in the boom and booming year, 1927; and the reputable New York State Commissioner for Labor stated that in 1928, with no corresponding slump in factory production, unemployment was as bad as it had been in 1921. Comparison of the years 1923 and 1927, by industries, gives overwhelming support to the conclusion.

Industry	*Change in Output*	*Change in Employment*
Oil: petroleum refining	84 percent more	5 percent less
Tobacco	53 percent more	13 percent less
Meat: slaughtering packing	20 percent more	19 percent less
Railroads, 1922–26	30 percent more	1 percent less
Construction, Ohio only	11 percent more	15 percent less
Automobiles, 1922–26	69 percent more	48 percent more
Rubber tires	28 percent more	7 percent more
Bituminous coal	4 percent more	15 percent less
Electricity, 1922–27	70 percent more	52 percent more
Steel	8 percent more	9 percent less
Cotton mills	3 percent more	13 percent less
Electrical equipment	10 percent more	6 percent less
Agriculture, 1920–25	10 percent more	5 percent less
Lumber	6 percent more	21 percent less
Men's clothing	1 percent more	7 percent less
Paper	0	7 percent less
Shoes	7 percent more	12 percent less

These figures have never, so far as I know, been attacked. Mr. Simeon Strunsky makes much of the fact that the census figures for 1920 and 1930 show an increase in the latter year of 1,280,000 in "manufacturing and mechanical industries," but neglects to explain how these occupation figures are obtained. As Mr. Strunsky knows, the decennial census is a population census and not an employment census. A man will speak of his "occupation" as machinist if he was last a machinist, even though he have now no job. And the tabulator has no choice but to write him down as a machinist. Thus the decennial census throws little light upon the problem. One must turn to the census of manufactures where, on the basis of pay rolls, actual employment figures are given. This census shows a drop from 9,000,000 in 1919 to 8,350,000 in 1927 and 8,839,000 in 1929.

Another device of Technocracy's critics is the use of percentages presented in Chapter VI of "Recent Social Trends," which show the proportion of the population, sixteen years of age or over, gainfully employed to have been 52.2 in 1870 and 57.1 in 1930. But this presentation is a trifle disingenuous since it omits the intervening figures. The peak shown by these percentages was not 1930 but 1910, when 59 percent were gainfully occupied. In 1920 the percentage fell to 58.1 and in 1930 to 57.1.

It seems impossible to deny the fact of the decline. And if the decline in number of wage-earners in the face of an increase in production is admitted, the case is proved. It would be no answer to argue, as many critics of Technocracy do argue, that the lost population of the factories was taken up in the "services"—the hotels and the bond offices and the gas stations. For a society which wished to take advantage of the potential benefits of the new industrialism would not force its displaced workers into these selling services to swell a new Coolidge-era expansion, but would decrease workers' hours to spread the leisure resulting from technological advance. It is only the present misfit distribution system which makes it necessary for one man to take to the road selling insurance while nine men left in the factory go on working ten hours a day. The "services" from this point of view are merely a buffer margin to enable the present system to frustrate its own genius in the interest of its creditors. And the benefits from swelling the services are, as the present disaster proves, temporary at best. But even were it otherwise, the opposition would still have failed to prove that the services will take up all the slack. Surveys have been made showing that 1,907,000 new positions were created in the twenties in medicine and hotels and restaurants and moving-picture theaters and banks and the like as against the 1,485,000 positions estimated to have suffered technological cancelation in industry. But this result neatly omits the population growth in the interval which should have increased the number of wage-earners by 2,000,000. And the only actual case studies so far made to determine how adequately men displaced as puddlers and cutters are really taken up in teaching and stock and bond offices have signally failed to prove the opposition point. The lag between jobs was found to be serious even when new jobs could be found at all, and the only conclusion to be drawn from these studies is that, however the individual components of the residuum of the unemployed may shift and change, the total body of unemployment remains. In the last analysis the opposition is left with the cold comfort of the orthodox economic theory that men displaced by the machine in one industry *must*, either because of reduced costs and greater demand or the deflection elsewhere of the consumer's money, find work eventually. That theory has never been proved.

If the opposition has failed to make its point on the basis of actual employment figures, it has also failed to make its point on the basis of the rate of change. The significant and fundamental fact which no one has denied is that productivity per wage-earner was almost constant for the twenty years from 1899 to 1919, whereas in the eight years following 1919 it increased by approximately *50 percent.* Critics have suggested that a similar or greater increase in productivity occurred at the beginning of the Industrial Revolution. But the eras are not comparable. Change from handicraft to machine production enormously increased output per worker and displaced thousands of men amid dire forebodings. But mechanization was still only an adjunct to human labor. As production increased with the manufacture of cheaper goods, employment also increased. Only now has man become an adjunct, and an increasingly less important adjunct, of the machine. Only in our time has an increase in production been possible with an actual decrease in number of men employed.

Technocracy as a group is far from invulnerable. Its statistics are shaky. Its utterances have been half-cocked. And its philosophy is merely an infantile doctrine of technological determinism to take the place of the equally infantile Marxian dogma of economic determinism. "All social activity," according to the Technocrats, "must obey the laws of physics." Human responsibility for human action is canceled and nothing is required of a man but that he should submit to the laws of physics, measure his life in ergs, and discard all interests which cannot be expressed in foot pounds per second. Technocracy, in other words, goes no farther on the philosophic side than to offer to an increasingly childish humanity another mechanical nurse, another external authority, and another absolute pattern. But for all that, the problem which the word Technocracy unfortunately defines is the vital problem of our time and the hope which the word Technocracy obscures is the first human hope industrialism has offered. Those who ignore the problem and those who discredit the hope do so at their peril.

CONCLUSION

The preceding selections cover a long span of American history (1832–1969), but persisting as well as changing attitudes can be found in them. The most pronounced and durable American attitude is the conception of technology as man's means of transforming the material environment. Thomas Ewbank wrote in 1855: "What then was it that was so conspicuously to mark his [man's] connection with the earth . . . it was the character he was to assume as a MANIPULATOR OF MATTER . . . the sounds of his implements acting upon it were to swell till their reverberations rolled over the whole." More than a century later both Lewis Mumford and Richard Buckminster Fuller believed that man's potential to transform the environment, natural and machine-made, was virtually limitless. A corollary of the persistent concept of man with his machines transforming the environment is the view of the world as an artifact or as "spaceship earth."

The vision of earth transformed into an artifact often appeared with a utopian cast: J. A. Etzler, whose book Thoreau reviewed in 1843, envisaged the man-made world as paradise regained. The implication that Eden, lost by the fall, would be regained through self-help, as it were, found place in an essay of 1832: "In the earliest ages of society, machinery was unknown. Man was created in a climate where the earth yielded bountifully at all seasons of the year . . . his only labor was to gather from Nature's abundant store, the supply of his present want." "The glory," the essayist continued, "of compelling the powers of Nature into the service of man, was destined to grace our own age." R. H. Thurston, writing almost sixty years later, anticipated that "we may be allowed to hope that later generations may continue to see an interminable succession of advances made by coming men of science, and by learned engineers and mechanics, that shall continually add to the sum of human happiness in this world. . . ." Within a few generations Thurston expected "a true millennial introduction."

The reader will also have observed that nineteenth-century

Americans expressing general views about the implications of technology for man focused on its labor-saving aspect. Understandably, the large segment of Americans who had known the exhausting, back-breaking toil of the countryside in the Old World and the New, or who had seen their parents and grandparents prematurely aged by labor, welcomed—even celebrated—the relief promised and delivered by labor-saving technology. The attitude toward the introduction of labor-saving machines was not altogether one of economic calculation. As will be discussed later, this attitude—understandably persistent in the nineteenth century—seems to have been less obvious in the twentieth. The nineteenth-century attitude toward labor and labor-saving machines also embraced the view that intellectual activity—in contrast to physical labor—was pleasant, dignified, and brought status. R. H. Thurston was so thoroughly convinced that increased intellectual activity and decreased physical labor were blessings of progress brought by applied science and engineering that he predicted with obvious satisfaction that the future would bring a change in man's physique: "Man reducing the powers of Nature still more completely to his service will depend less on the exertions of his muscles, and they will be correspondingly and comparatively less powerful. . . ." "The brain will be developed," he anticipated, "to meet the more complex and serious taxation of a more complex and trying civilization. . . ."

Understandably, man in becoming more intellectual, mastering nature, and making a new environment would take on creative—even godlike—characteristics. This attitude emerged clearly in the earlier years, but tended to sour in the 1960s, for as the character of the handiwork became apparent, the maker became suspect. Early in the nineteenth century, Thomas Ewbank asked rhetorically what else man could do on earth but imitate "the Great Artificer of all that moves." In 1896 the exhilaration of creativity still persisted so keenly that Edward W. Byrn could exclaim: "The old word of creation is, that God breathed into the clay the breath of life. In the new world of invention mind has breathed into matter, and a new and expanding creation unfolds itself." Joseph K. Hart wrote a generation later: "He [man] must make nature over, on lines nearer to his needs; he must turn nature into a great machine, subject to his control, serving his every desire. This was an adventure so presumptuous and so perilous that most subsequent ages have called it irreverent, profane. Some have even spoken of it as the 'Fall of Man.' "

By the end of the nineteenth century, Americans had come to see themselves as uniquely and outstandingly creative and their technology as unusually effective. If any one people should be singled out for their godlike aspirations and pretensions for doing over what had already been done in the beginning, it was the Americans. Edward W. Byrn rhetorically asked why most of the great technological developments of the late nineteenth century were of "American authorship," and replied that beneficent institutions and laws that recognized the inventor as a public benefactor were the explanation. Appraising American inventive genius as it faced the challenge of national defense, the Secretary of the Navy expected to strike a responsive chord as he asserted: "American genius will meet any needs and meet them as they have not been met by the conventional methods of the older nations." Despite the technological unemployment of the Great Depression, Howard Scott, the spokesman for Technocracy, believed America so far ahead of Europe in industrial development that she could learn nothing from the Old World, but had to find her own way into a new era of fully utilized engineering genius and technological power. By 1900 the American attitude toward technology reflected an awareness that the country had emerged as the world's leading industrial and technological power when its achievements were measured by such conventional yardsticks as production statistics for heavy industry, transportation facilities, power generation, and patent activity. For the next half-century, shaped by this knowledge, the attitude persisted.

Another recurring twentieth-century attitude stemmed from the realization that technology should be conceived of as an abstract force applicable to a variety of problems. Earlier there was a concentration—almost a fixation—upon a few specific problems like the making of a hospitable environment from a hostile wilderness, the relieving of the burdens and shortages of labor, and the reinforcing of democracy through the better distribution of goods and services. Perhaps Henry Adams expressed most imaginatively the more recent and more flexible concept of technology as an abstract force. To Adams, the dynamo, or electrical generator, was a symbol expressive of ultimate energy; he "began to feel the forty-foot dynamos as a moral force, much as the early Christians felt the Cross. . . . Before the end one began to pray to it, inherited instinct taught the natural expression of man before silent and infinite force." Joseph K. Hart, an articulate social critic of the 1920s, believed that "we stand today where the Greeks once stood: face to face with Fate. We have Power beyond their dreams of power. . . ." An outstanding American physicist, Robert A. Millikan, contemplating science and modern life, concluded in 1928 that the modern world was distinctive because of "the discovery of the very idea of progress, for the discovery of the method by which progress comes about, and for inspiring the world with confidence in the values of that method." Applied science and technology had been transformed in the minds of many Americans from concrete machines and processes with narrow functions into abstract forces with undefined—even unlimited—potential.

This lifting of technology to the level of generalized power or force raised the problem of decision and control, a problem that escaped the notice of many early nineteenth-century Americans. They seem to have assumed that benevolent providence, and/or a law of beneficent progress, determined the unfolding of technology in relationship to American society. Hindsight suggests that harnessing natural power and cultivating the land were such pressing and clearly desirable applications of technology that comparable alternatives were not explored. Hence the nineteenth-century feeling that technology was unfolding without need for human planning and design on a grand scale.

Perry Miller wrote that early nineteenth-century Americans "flung themselves into the technological torrent"; they shouted with glee "and cried to each other as they went headlong down the chute that here was their destiny. . . ." "The age was grasping for the technological future, panting for it, crying for it." Writing of the 1960s, Lewis Mumford used a different metaphor, likening his age to a driverless automobile, filled with passengers, "hurtling full speed toward doom." The metaphors are generally similar conceits of passengers hurtling into the future, but with a striking difference: Miller's nineteenth-century Americans innocently gave themselves up to the sweep of technology, joyously embracing a vision of the technological age; Mumford's twentieth-century passengers see themselves as unwilling occupants of a vehicle gone out of control. This change in attitude has been one of the most dramatic in history.

Only a minority of writers—albeit ones well read and well regarded in American literature courses—protested in the nineteenth century against the majority's infatuation with the machine and the prospect of a technological age. Thoreau did protest against the easy assumption that such an environment, even if made to man's specification, would succor the inner spirit ("When the sunshine falls on the path of the poet, he enjoys all those pure benefits and pleasures which the arts slowly and partially realize from age to age"). Yet, it was not until World War I shattered the Victorian optimism and complacency that anyone began to question whether technology might not be carrying man to a benevolently foreordained destiny. The euphoric drift of a century had been stopped. The shift to anxiety and a sort of despair, however, was gradual; before men like Mumford abandoned hope, they saw a real possibility—to extend the metaphor—of moving into the driver's seat.

The possibility of assuming control, of interrupting the rush and momentum of technology is inherent in the essay by Bertrand Russell, in the articles from *The Literary Digest* about the need for an applied science moratorium and for disarmament, and in the article by Joseph K. Hart. All three were written in the interwar period of disillusionment and ambiguity. Russell believed that the nineteenth century had been a fool's paradise which nurtured the anticipation that invention after invention was orderly progress. The illusion collapsed, Russell noted, in the slaughter of 1914. He lamented man's failure to control technology, but he did not eliminate the possibility that such control could be imposed. Russell's metaphor was a youth in a raccoon coat driving a sixty-horsepower Rolls Royce "with a dozen cocktails inside him." There is, however, someone in the driver's seat, and he might be restrained from drinking. Joseph K. Hart shared Russell's pessimism concerning the past and probable social effects of applied science and technology, but Hart also saw the possibility of man asserting his will, developing new technology, and solving the unfortunate legacy of past science and technology. The assumption that man could and should assert control was also manifest in the Bishop of Ripon's call for a moratorium.

Having seen the possibility and need for man's directing technology on a grand scale, some Americans, so persuaded, took the next logical step and formed opinions about who should direct or control. Essays written between the wars recommended that scientists take the responsibility. Arthur D. Little, an industrial chemist and advocate of industrial research, supposed that scientists and engineers were clear-headed, well-informed, rational, and most important, that the scientific method was generally applicable to society's problems. Little believed that a moderate reorganization of the political structure resulting in increased utilization of scientists in government would be a means of avoiding in the future the shocks and disappointments of the twentieth-century world of science and technology. Dynamite, TNT, poison gas, airplanes, and motor cars should not be controlled by criminal or ignorant types, but by scientists, who *ipso facto,* Little believed, were enlightened men of good will.

Robert Millikan, a Nobel Laureate in physics, thought that "the immediate destinies of the race" were in "our own hands." Like Little, Millikan was convinced that scientists should play a prime role, for they understood best the scientific method, "the only hope of the race of ultimately getting out of the jungles." Compared to earlier and lesser civilizations the modern world was distinctive, Millikan had decided, because of its discovery "of the very idea of progress, for the discovery of the method by which progress comes about." His confidence was strong, for "so long as the world can be kept thus inspired, it is difficult to see how a relapse to another dark age can take place." Arthur Compton, another Nobel physicist, expressed a more guarded optimism. Like Millikan, he advocated that scientists use the rational scientific method to solve the problems of the new world structured by science and technology.

Early in the 1930s others appealing to a general audience of Americans asked that engineers and industrial managers assume responsibility for organizing and administering modern technology and science-based society. While the proposals of Compton, Millikan, and Little advocated that scientists play a more influential role within the existing political and social structure, Technocrats, proponents of broad decision-making power for the engineers and industrial managers, demanded a thoroughgoing reorganization of society. They too insisted upon a rational program that would bring the clarity, logic, and empiricism of the engineering mind to bear on social problems. A fundamental postulate of the Technocrats was that the Great Depression had been caused not by technology but by the irrational mismanagement of technology by bankers, politicians, and businessmen. The commitment of engineers to rational organization of the means of production and of related societal functions would bring abundant production and a managerial meritocracy based on competence.

Although seemingly a flash in the pan by the mid-thirties, technocracy survived a generation later. Theodore Roszak, writing in 1969, felt that "our society advances toward technocracy along many paths." Roszak was referring not to the moribund Technocracy movement of the early thirties, but to the essential ideas of that movement. His attitude toward it was hostile; he characterized it as "a massive historical movement whose seriousness we underestimate at the peril of our human dignity." He wanted technology brought under control, but certainly not by engineers and industrial managers.

Judging by the considerable interest in Roszak's views expressed in the late sixties, his attitude struck a responsive chord. He lamented the Technocrat's single-minded and myopic dedication to rational and scientific methods. One manifestation of the engineering approach in 1969 that Roszak found particularly offensive was the "objective" approach of the systems analyst. Reserving his sharpest criticism for R. Buckminster Fuller, Roszak characterized him as "a great bamboozler: a combination of Buck Rogers and Horatio Alger." Interestingly, Fuller's program for ordering "spaceship earth" was reminiscent of the writings of Ewbank, Etzler, and other early nineteenth-century Americans determined to use technology to create a new world. Expressing and formulating American attitudes a century later, Roszak wrote feelingly of sky, wind, clouds, trees, water, and other non-man-made things. The contrasts between Roszak, the Technocrats, and scientism show clearly the conflicts and ambivalence in American attitudes.

A similar change in attitude can also be identified by comparing the recent writings of Lewis Mumford and the earlier ones of Millikan and others enthusiastically advocating the general applicability of the scientific method. Mumford lamented the rise of crass mechanistic notions that value efficiency above all. Simplistic nineteenth-century utilitarianism survived, Mumford asserted, in the twentieth-century "guises of technocracy and scientism." Elsewhere, he wrote an impassioned attack on the reductionism of the scientific method triumphant in the West since Galileo. Its preoccupation with primary quantities—the measurable and the manipulable—contributed, Mumford had perceived, to the organization of life devoid of qualities and activities like "ritual, art, poesy, drama, music, dance, philosophy, science, myth, [and] religion." Mumford did not totally reject science and technology; he only abhorred their "overgrowth" at the expense "of human needs and aspirations."

Having noted the shifting attitude toward who or what should be held responsible for remaking the world, one can consider changing American attitudes toward their handiwork. A nineteenth-century enthusiast asserted joyously, "Vanquished Nature yields. Her secrets are extorted. Art prevails." A century later, Americans believed that they had befouled their nest. The basic problems of a hostile—or at least indifferent—natural environment had been solved. The relative freedom in twentieth-century America from attacks by wild animals, the effects of extreme heat or cold, the devastation of floods, and the privations of drought and plague gave evidence of remarkable achievement. Ironically, however, the hostile forces of a natural world had been replaced by the hostile forces of the man-made or artificial one. Wild animals had their counterparts in the irrationality of muggers and rapists driven to their deeds, many believed, by the deep distress of the urban environment; the extremes of heat and cold had equivalent health hazards in smog alerts; the devastation of floods had an equivalent in the uncontrolled and deadly flow of onrushing traffic and a spreading highway network that eradicated the immediate environment. The parallels were numerous, but it took several centuries for many Americans to realize that the artificial world might be as hostile—at least as indifferent—as the natural one so ruthlessly subdued.

"In the last few years," Barry Commoner observed, "with a sudden shock, it has become apparent that modern technology is changing the environment—for the worse." The essay by Commoner manifested the delayed—and in his case, penetrating and comprehensive—realization that art was not always an improvement on nature. Commoner chastized Americans for having tried to subdue nature rather than live in it. (He might have cited as evidence such exhortations as Thomas Ewbank's in 1855 that man manipulate matter, imitate "the Great Artificer," and make the world an artifact. Commoner suggested subtle adjustments of man, nature, and technology. As a scientist aware of the infinite complexities of the mix of the man-made and natural in the biosphere, Commoner urged restraints far more sophisticated than a lover of nature, like Thoreau, could have conceived of a century earlier.

American attitudes toward the man-made world and the role of man in general, and of scientists and engineers in particular, in the creation of this world have changed over time. Other changing attitudes can be associated with Americans' views of technology in relationship to national goals. The early selections by "T" (Thoreau), Ewbank, and the reviewer of *The Results of Machinery* focused on the work that needed to be done to cultivate the wilderness. In a selection written in 1855, Denis Olmsted insisted that Americans' determination to establish a democratic society was reinforced by the applications of science and technology. The inventiveness of Americans during the half-century before 1896 proved to "Beta" (Edward W. Byrn) that the nation had emerged as the eminently civilized and resourceful Western power and that inventiveness would allow it to remain so. He believed that inventiveness and the technology flowing from it led the nation progressively to a utopian future. The engineer and educator R. H. Thurston could only assent.

Technology was also considered a means of enduring national defense and of winning wars, as Thomas A. Edison and Secretary of the Navy Josephus Daniels testified during World War I. Others were sure that science and technology could be used to promote disarmament, if the likelihood that modern war would wreak terrible destruction upon both victor and vanquished was understood. If depression came, then technology—rationally exploited by a rationally organized society—could reverse the trend. These shifts suggest that technology has been chameleon-like, changing character to fulfill national goals. And there is no reason to doubt that these shifts will continue—even the strongly held views of a Mumford or a Roszak may tomorrow appear ephemeral.

Another characteristic of American attitudes revealed in the sample taken in this book has been the inherent fallacies they contained. The early forecasts embodied in these attitudes seem incredible in view of later history. The euphoric essays of Ewbank, Etzler, Thurston, and "Beta" could not be taken seriously today as comments upon technology and society, even if the anachronisms were excised and the style altered. Their projections of the future were simplistic and uncritical; they had no concept of unwanted secondary effects (such as smog and slums) flowing from the intended ones (industrialization). The nineteenth-century attitude expressed by Thurston and the others was suggestive, as Bertrand Russell observed, of a fool's paradise. The most obvious explanation for their naiveté is their assumption that technological developments were governed by an orderly law of progress. So assured, they called for no general planning, no "technological assessment," to use recent terminology.

The reviewer of *The Results of Machinery* could not conceive of serious technological unemployment in America, nor did he foresee that labor-saving machinery could lead to boring, repetitive work. Olmsted believed that closer communication through new transportation and communication technology would bring mutual understanding and respect. "Beta" had no doubt that civilization (high culture) and mechanical invention were and would be correlated, "invention the cause and civilization the effect." In 1924, *The Literary Digest* took seriously scientists who believed that the fear of more destructive weapons would bring disarmament and make the very thought of war intolerable. A. D. Little contended that the "fifth estate" would prove peculiarly benevolent; Robert Millikan thought the scientific method would prove generally applicable. An American can take only a perverse satisfaction in the fact that the illusions and delusions were not peculiarly American; he can only hope that growing realization of the complex character of technology, resulting from having seen it unfold in history and from having realized the fallacies in men's attitudes toward it, will bring Americans to insist on fuller historical understanding, informed policies, and critical attitudes.

75 76 77 9 8 7 6 5 4 3 2 1